WRECKCHASING

A Guide To Finding Aircraft Crash Sites

By Nicholas A. Veronico

With Ed Davies
A. Kevin Grantham
Robert A. Kropp
Enrico Masagali
Thomas Wm. McGarry
Walt Wentz

Dedicated to the pilots, air crews, and their families whose loved ones perished. May their crash sites be found so that their sacrifices may be remembered.

Published by Pacific Aero Press
P.O. Box 20092
Castro Valley, Calif. 94546-8092

ISBN 0-9636332-0-1
First Edition April 1993
Printed and bound in the United States of America

All information contained in this volume is accurate at the time of publication. No guarantee is made to the existence, location, or condition of any aircraft wreck. This volume is intended to be a guide for research purposes only.

Type by Third Avenue Graphics
2157 Fleur de Lis Court
Arlington, Texas 76012

On the Front Cover:

During a routine low-level training flight from NAS Moffett Field, Calif., to NAS Fallon, Nev., two Douglas AD-6 Skyraiders from Navy Attack Squadron 195 crashed into the side of Mt. St. Helena at the north end of California's Napa Valley. Weather for the day was low fog and light rain. AD-6, Bureau numbers 135254 (shown) and 135264 struck the 4,344-foot Mt. St. Helena at the 3,800-foot level. The tails of both aircraft are resting about 20 feet apart at the 2,800-foot level. The pilots were Lt. Cmdr. Robert L. Worrell and Lt. (j.g.) Roger W. Tillson Jr.

Photo by Nick Veronico

On the Back Cover:

United Airlines DC-3A, N16066, crashed on the beach at Point Reyes, Calif., at 5:30 a.m. on November 29, 1938. The DC-3 was en route to San Diego from Seattle, Wash., via Portland and San Francisco.

Photos via William T. Larkins Collection

CONTENTS

ACKNOWLEDGMENTS

This guide to finding aircraft wrecks came about more as a hobby than anything else. Along the way, I have received help from and made lasting friendships with many people who share this love. Thank you all.

A special thanks to my research and hiking partner Ed Davies. He, along with Chris Kopel and Gary Larkins, got the Wreck Chasing bug started in me.

Thanks are due to the following for their unselfish time, energy and written contributions: Thomas Wm. McGarry, A. Kevin Grantham, Robert A. Kropp, Enrico Masagali, and Walt Wentz, who provided a number of stories to help round out this volume. Gary Larkins provided a large number of photos.

Preparation of the Philippine Clipper chapter was assisted by Bo Hiatt, Gary Urdahl, Elgin Long, Donna and Larry Struck, Gordon Werne, and Lila Lee of the Mendocino County Historical Society.

The BCPA chapter started with a conversation between Dick Krug and Joel Harris, plus six slides. Research materials and guidance were provided by Reggie McGovern, Ray Brooks, John Escobar of the Mid-Peninsula Regional Open Space District, Mitch Postel of the San Mateo County Historical Society, Ken Fisher, Bob Swanson, and Roy Myers.

A number of others are owed a debt of gratitude for their assistance in providing access to wreck sites and research materials. Thanks to: Nancy and Keith Hurt who showed the way to the Covelo B-17; Murray Moore; Tighe Ramey; Patrick Congdon — Mid-Peninsula Regional Open Space District; Gary Copeland and Betty Anderson of Pacific Photo; Clay Jansson; Maj. John Elliott, U.S. Marine Corps, ret., of the Aviation Branch, U.S. Naval Historical Center, Washington D.C.; John Shackleton and Anne McMillin from the Public Affairs office at NAS Moffett Field; Sally Gersbach; Veralyn Davids; John Harjo; Ron Strong; Scott Thompson; William T. Larkins; Jerry Knight; Sig Unander; Pamela Veronico; Armand Veronico; Karen Haack; John Wegg; Jay Wisler; Vickie and Ciro Buonocore of *In Flight Aviation News*; Michael A. Bodisco — Bureau of Alcohol, Tobacco and Firearms; Greg Stone and Harrison Rued; plus many of the members of the Redwood Empire Aviation Historical Society.

PREFACE

Aviation holds a certain intrigue in people's minds. Everyone has heard the stories of airplane wrecks. They are much like the tales of sunken ships. Wreck Chasing provides the thrill of discovery mixed with the romance of flight.

This book is the tool I never had. Trial and error, dead ends, and hours of picking other people's brains led to the wrecks I have visited. There is nothing better than a good day outdoors. Of course, there is nothing better than stepping out of your car and walking 50 feet to an aircraft wreck, but for me that has not been the case.

I have confined my searches primarily to military aircraft and airliners that crashed prior to 1959. In my experience, almost all of the civilian aircraft wrecks have been removed either by their owners or their insurance companies. Military aircraft, which crashed in the 1940s and 1950s, went down in areas that were inaccessible at the time. With urban expansion and the logging of our forests, roads now crisscross these once remote areas. But unless you are specifically looking for a crash site, anyone traveling down these roads would not even see wreckage of a plane lurking nearby.

In almost all military wrecks from the 1940s, the Army Air Forces or the Navy visited the site, removed the remains of the crew, and recovered or destroyed the bombsights, avionics and radio gear. Rarely was the armament removed or the site cleared of wreckage. With this in mind, I have focused on military wrecks as I feel there is a much better chance of finding an aircraft's remains.

In the list of aircraft wrecks, I have provided the longitude and latitude. This is a starting point. These areas are general in nature, but they are the known crash points of aircraft. The reader must undertake further research, as outlined in Part I, to find an aircraft's exact location.

I have also included a number of stories of aircraft wrecks that readers of this book can locate easily. I hope you will consider using these sites as a starting point in your search. Also included are stories of aircraft recovered by the military and by professional aircraft salvage companies. Many aircraft restorers use components from these wrecks for patterns to replace parts too worn or unavailable from any other source.

Also, this book should provide you with the names and addresses of a number of helpful information sources.

This book is intended as a starting point for those looking to find aircraft wrecks.

This TBF is a graphic display of just how fine the line between life and death is. This Avenger missed the mountaintop by only 50 feet.
Photo by Gary Larkins

PART I

Wreck chasing: locating wrecks

Since the dawn of time, man has envied the bird. After the Wright Brothers took wing at Kitty Hawk in 1903, man has been obsessed with flight. The airplane was first a curiosity — only for the brave; then a military weapon; later a pleasure craft; and most recently, a way to shrink the earth. Air travel has brought us to within three hours of Europe and only a day's flying time to Australia or the Great Pyramids of Egypt.

What goes up must come down. Ninety-nine point ninety-nine percent of the time aircraft fly without incident. It is that .01 percent that intrigues everyone. When an airplane crashes, it is headline news, and television news loves aircraft crashes. A non-fatal crash of a single-engine Piper or Cessna will be shown on every news promo during the evening in an attempt to get you to watch that station's late-night newscast. Some crashes have garnered almost as much air time as the sportscast.

The older the wreck, the better the chances of its remaining where it crashed. With the exception of general aviation airplanes, aircraft wrecks prior to 1960 are usually right where they landed.

Getting Started

What is the first step to finding an aircraft crash site? There are two ways to begin your search. One way is to find someone who has been involved, either as an eyewitness, an on-scene police officer or firefighter, a volunteer aiding in the rescue effort, or just a local who ventured to the sight months after the crash occurred. If you do not know any of these people, how do you find them?

The second way is to obtain a "crash locator" list. Almost 500 known aircraft wrecks are listed in Appendix I, which contains the wreck sites from two such lists. Of course, there are many aircraft out there that are not on the list, but this is a good starting point if you have not met someone who has first-hand knowledge of an aircraft crash site. The wreck locators usually provide an aircraft type, location by longitude and latitude, date of accident, and any remarks, such as aircraft condition or nearest topographical landmark. These "exact locations" are provided for aerial search teams looking for lost aircraft. If wreckage is spotted, it can be checked against known wrecks so that valuable search team efforts are not spent hiking to a 40-year-old wreck site.

Doing Your Homework

Even if you have been told the "exact location" of a wreck or know an eyewitness, head to a public or college library to begin your search. If you do not have an aircraft crash you are already researching, decide on what era you want to cover. The period from the start of World War II through the end of the Korean War, 1942 to 1953, will be the most fruitful.

Begin with a local newspaper published in the area of a crash. Back copies will be available in a bound format or on microfilm. Microfilm is best as you can quickly scan each day and then move on; plus, for a small fee, you can have a hard copy of any particular day. Note that if an aircraft crashed on the other side of the country and was carrying any hometown crew members, it will be reported in the local paper. If you are looking for a wreck from the wreck locator and know the date, your newspaper search will be brief, but the information invaluable.

Five Killed in Crash of Bomber Near San Mateo

A four-motored army bomber crashed last night at Native Son's Park, on the King Mountain road, San Mateo county, carrying to death its five occupants.

Hamilton Field public relations officers said late last night the plane was on a combat training mission from Boise, Idaho, to Oakland.

The plane crashed in a heavy drizzle and burst into flames.

The bodies were recovered by sheriff's deputies. Names of the dead were withheld pending notification of next of kin.

The plane earlier had been reported missing. It had not issued a report since leaving its clearance post not far from its California destination, a Fourth Air Force public relations officer said.

One of the many tragic wartime aircraft crashes that appeared in the San Francisco Chronicle *December 25, 1943. This tidbit of information is the beginning of a Wreck Chase.*
Ed Davies Collection

The newspaper will give you a number of key bits of information: the date of the accident — very important for obtaining official reports and records; the type of aircraft; who was involved — those souls on the airplane; and, most importantly, the names of the local people and safety agencies who participated in the rescue or the crash investigation, as well as the condition of the wreck. Be sure to scan the following two weeks or so to find out if the wreckage was removed.

Most of all, the newspaper reports will tell you in understandable terms where the wreckage is located. Directions from the newspaper usually read, "The wreckage of the P-38 is one-quarter mile southeast of the Vista Point off Redwood Road in the trees bordering the Johnson Ranch...," while the location of aircraft wreckage from an official report is, "crash occurred at Latitude 39 04'N, Longitude 123 17'W... ." This area could encompass up to five square miles, thus the importance of finding an account of the crash in the newspaper.

Interviewing The Participants

Once you have learned the identity of the aircraft you are searching for and the names of those who participated in the crash's original investigation, you are ready to move forward. Taking the names reported in the paper, contact the newspaper to see if the reporter who wrote the story or the photographer are still in the area. The reporter will know the exact location and may have insights and details of the aftermath of the crash — some of which may not have made it into print. The newspaper may have a photo archives or library that still holds the negatives from the crash, or the photographer may have them.

The reporter, photographer, newspaper archivist, and librarian are excellent people to interview. After the reporter and photographer, start with the phone book. You will be amazed how many "old timers" are still around. Most of them have excellent memories, and the sight and story of a plane crash are not likely to be forgotten.

The crash of the British Commonwealth Pacific Airlines Douglas DC-6 on October 29, 1953, killing 19, involved many agencies. The county sheriff was involved along with a number of local police and fire agencies, as well as the American Red Cross. While looking for this wreck, we used the newspaper account and cross-referenced it in our search of the telephone directory, which revealed the names of a number of police and fire officials, Red Cross volunteers, as well as the dentist who identified the bodies during the post-mortem examinations. As we contacted each person, each remembered the wreck and was able to provide us with the location, the easiest way to get to it, and what geographical landmarks to look for, as

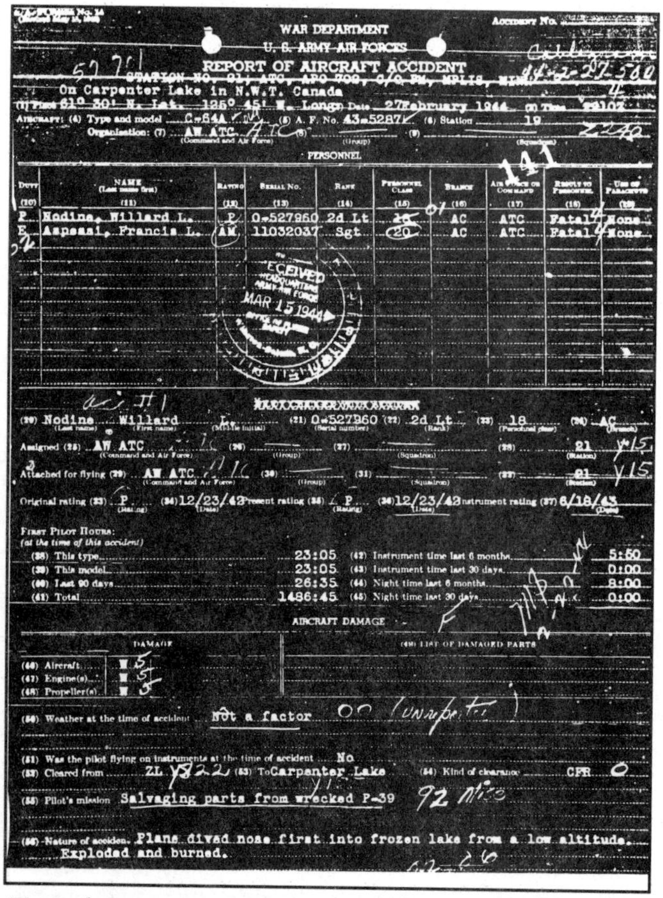

WAR DEPARTMENT
U. S. ARMY AIR FORCES
REPORT OF AIRCRAFT ACCIDENT

On Carpenter Lake in N.W.T. Canada
61° 30' N. Lat. 125° 45' W. Long. Date 27February 1944
Aircraft: (4) Type and model C-64A (5) A. F. No. 43-5287 (6) Station 19
Organization: (7) AW ATC

PERSONNEL

Duty (10)	Name (11)	Rating (12)	Serial No. (13)	Rank (14)	Personnel Class (15)	Branch (16)	Air Force or Command (17)	Result to Personnel (18)	Use of Parachute (19)
P	Nodine, Willard L.	P	0-527960	2d Lt		AC	ATC	Fatal	None
E	Aspeasi, Francis L.	AM	11032037	Sgt	20	AC	ATC	Fatal	None

RECEIVED MAR 15 1944

(20) Nodine Willard L. (21) 0-527960 (22) 2d Lt (23) 18 (24) AC
Assigned (25) AW ATC (28) 21
Attached for flying (29) AW ATC (32) 21
Original rating (33) P (34) 12/23/42 Present rating (35) P (36) 12/23/42 Instrument rating (37) 6/18/43

First Pilot Hours: (at the time of this accident)
(38) This type 23:05 (42) Instrument time last 6 months 5:50
(39) This model 23:05 (43) Instrument time last 30 days 0:00
(40) Last 90 days 26:35 (44) Night time last 6 months 8:00
(41) Total 1486:45 (45) Night time last 30 days 0:00

AIRCRAFT DAMAGE
Damage: (46) Aircraft W (47) Engine(s) W (48) Propeller(s) W
(49) List of damaged parts

(50) Weather at the time of accident Not a factor
(51) Was the pilot flying on instruments at the time of accident No
(52) Cleared from ZL (53) To Carpenter Lake (54) Kind of clearance CFR
(55) Pilot's mission Salvaging parts from wrecked P-39
(56) Nature of accident. Plane dived nose first into frozen lake from a low altitude. Exploded and burned.

Typical Army Air Forces crash report. Page One gives the rough statistics, such as aircraft type, crew names, weather and flight plan.

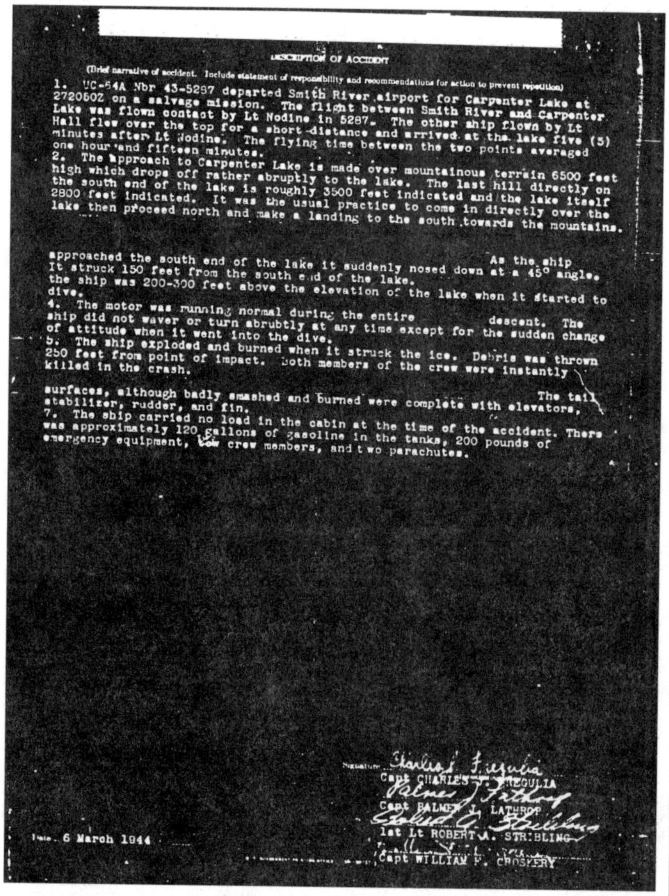

DESCRIPTION OF ACCIDENT
(Brief narrative of accident. Include statement of responsibility and recommendations for action to prevent repetition)

1. UC-64A Nbr 43-5287 departed Smith River airport for Carpenter Lake at 272050Z on a salvage mission. The flight between Smith River and Carpenter Lake was flown contact by Lt Nodine in 5287. The other ship flown by Lt Hall flew over the top for a short distance and arrived at the lake five (5) minutes after Lt Nodine. The flying time between the two points averaged one hour and fifteen minutes.
2. The approach to Carpenter Lake is made over mountainous terrain 6500 feet high which drops off rather abruptly to the lake. The last hill directly on the south end of the lake is roughly 3500 feet indicated and the lake itself 2800 feet indicated. It was the usual practice to come in directly over the lake then proceed north and make a landing to the south towards the mountains.

approached the south end of the lake it suddenly nosed down at a 45° angle. As the ship It struck 150 feet from the south end of the lake.
the ship was 200-300 feet above the elevation of the lake when it started to dive.
4. The motor was running normal during the entire descent. The ship did not waver or turn abruptly at any time except for the sudden change of attitude when it went into the dive.
5. The ship exploded and burned when it struck the ice. Debris was thrown 250 feet from point of impact. Both members of the crew were instantly killed in the crash.

surfaces, although badly smashed and burned were complete with elevators, The tail stabilizer, rudder, and fin.
7. The ship carried no load in the cabin at the time of the accident. There was approximately 120 gallons of gasoline in the tanks, 200 pounds of emergency equipment, crew members, and two parachutes.

Date 6 March 1944

Capt CHARLES J. TREGULIA
Capt BALMER J. LATHROP
1st Lt ROBERT A. STRIBLING
Capt WILLIAM F. CROSKERY

Page Two is a narrative of the crash signed by the reviewing officers.

Gary Larkins Collection

well as many unreported details. The firsthand accounts of the crash scene were unbelievably interesting.

The county historical society and the city historian are two excellent places to look for information. If they cannot help, they can always provide you with names of locals who might have been involved or the name of the landowner — past and present — of the property where the crash occurred.

If the crash involved a fatality, check the records at the county coroner's office. The pilot's and passengers' death certificates will be held either with the coroner or at the county's Hall of Records. The coroner's records will provide you with the name of the pilot and the others involved, location of crash, time of death, and next of kin. Again, the records are filed by date.

Many of the people you contact will be very helpful; others will be negative. While searching for a bomber wreck, a gentleman — who we later found out had never been to the site — told us that, "Hippies had gone in there and took out all the aluminum and sold it for scrap in the '70s." It just wasn't so. First of all, there probably was not a road near the crash site until the 1980s, and the site was so far from the nearest town that it would have been uneconomical for anyone to remove the aluminum. You've got to weed through the nay-sayers and B.S.ers.

Documentation Of Aircraft Crash

By now you have settled on a crash and probably know the aircraft type and date of the crash. The following government agencies can provide the remaining pieces of the puzzle or add some very interesting supporting documentation to your aircraft crash research.

You have chosen your wreck, researched the newspaper accounts of the crash, interviewed the participants in the rescue/recovery, and determined the wreck's general location. Before starting out on your trek, make sure you obtain the U.S. Geological Survey map for the area. See Appendix IV, Obtaining The Right USGS Map.

Where To Obtain Documentation

If you are searching for a civil aircraft and know its identity, you can obtain the crash reports from the National Transportation Safety Board. The date and location of the crash will be required. If you know the aircraft's serial number and civil registration, also enclose that information in your request. Write to:

National Transportation Safety Board
800 Independence Ave. S.W.
Washington, D.C. 20594

For all military aircraft, the crash reports are held at the Air Force Inspection and Safety Center. Again, you will need the date, location and the aircraft type, plus the serial number — if available. It is an Air Force policy to censor all reports before they are distributed to the public to protect the confidentiality of the crash investigators' opinions. The censor marks do not interfere with the location information. Forward your requests for military aircraft crash reports to:

Department of the Air Force
Headquarters Air Force Inspection & Safety Center
Norton Air Force Base, Calif. 92409-7001

After 8/93 — Department of the Air Force
Headquarters Air Force Inspection & Safety Center
Kirtland Air Force Base, N.M. 87117

Please note that both of the above-mentioned agencies require that your request be in writing. They do not have the staff to process telephone requests.

If you are looking for one specific type of aircraft, there are two ways to research a particular type. Each aircraft in the U.S. Navy or Air Force inventory is assigned a serial number and a corresponding record card that details the aircraft's assignments. Once you have settled on a particular type of aircraft, determine the serial number range assigned to that model. Most hobby shops carry a large selection of books that detail this information, or you may obtain a copy of U.S. Military Aircraft Designations and Serials Since 1909 compiled by John M. Andrade.

To order individual aircraft record cards by serial number, forward your request to:

U.S. Navy Aircraft
Department of the Navy
Naval Aviation History and Archives
Building 157-1
Washington Navy Yard
Washington, D.C. 20374-5059

U.S. Air Force Aircraft
United States Air Force
Historical Research Center
USAFHRC/HD
Maxwell Air Force Base, Ala.
36112-6678

To order aircraft record cards on microfilm, consult Appendices II and III.

If your interest is primarily U.S. Navy aircraft, you really are in luck. The Navy maintains a set of microfilm reels detailing aircraft accidents and incidents by aircraft type and date. The reels, known as Pilot History Cards of Aircraft Accidents, are $10 each, but well worth it. See Appendix II, U.S. Navy Pilot History Cards.

Turret cover from a Grumman 150-CE upper gun turret on a TBM/TBF torpedo bomber. Photo by Gary Larkins

In The Field

After at least one-half hour of digging and pulling, Patrick Congdon retrieved this generator motor from an F2H-4 Banshee. It was too heavy to carry out.
Photo by Nick Veronico

Engine, cowling and prop at the crash site of a Grumman F6F-5N Hellcat night fighter at Kahoolwe, Hawaii. Photo by Gary Larkins

Heading to the wreck site

Before heading to a site, determine who owns the land where the aircraft crashed. If the land is privately owned, contact the owner and obtain permission to venture onto the property. If possible, have the owner accompany you to avoid any problems. Otherwise, obtain written permission from the owner and carry the document with you. If the aircraft wreck is on public lands, which many are, check with the local governing authority. Determine the times and conditions of land use, locations of roads nearest your destination and their condition, and whether any locked gates or barricades will restrict your entry. Obtain and carry with you a topographical map of the area you intend to search. Having the map handy will keep you from getting lost and will aid you in subsequent trips. (See Appendix IV, Obtaining The Right USGS Map.)

Once you have selected a day to visit the site, gather your research and plan your route. You should never go alone, so invite a couple of friends to join you. Of course, if you don't find the wreck on the first day, they may become your ex-friends.

You'll need your USGS topographical map; a compass; good shoes; a pack with food and water; a lightweight jacket for traveling through brush; eye protection, such as sunglasses or safety glasses; a small digging tool; and a camera, flash and plenty of film. Also consider taking toilet paper, sun screen, and small plastic bags for parts you may pick up. This list of supplies should not have to be mentioned, as most people devoted enough to go looking for a wreck have the common sense to think of these things beforehand. However, on a recent trek, four of the eight explorers who set out for the crash site were unprepared. Have a good time, but remember, safety comes first.

Eureka, You've Found It

When you locate the wreckage, take a minute to focus on what is in front of you — a point of impact, a debris field, lots of scattered metal, and most of all, the exact spot where an aircrew met its end. I recommend sketching the wreckage quickly, making notes as to the terrain and where the wings, fuselage, engines and other large pieces of wreckage are located.

Do some of your own "accident investigation." Are there trees that have been topped? Did the airplane strike those trees during its descent before crashing? Are there major parts, such as wings or an elevator, more than half a mile away? Did the aircraft suffer a structural failure or did it enter a high-speed dive and shed those pieces on its way to the ground?

You will always be able to discern the wings, tail and any other large intact pieces. But what of the thousands of smaller pieces? Look for part numbers stamped on the formers and other components, data plates, numbers painted on the surfaces, overhaul tags, etc.

If you turn a piece of metal over and it reveals a painted surface, turn it back, face down, after you have studied it to preserve it for those who will follow. It is amazing how quickly the paint will fade when exposed to the elements.

After examining painted surfaces, turn them face down to preserve the paint for the next wreck chasers. Here the author examines the fuselage star and bar on a B-17F.

Photo by Murray Moore

Locate the point of impact. Check the surface for interesting parts, then kneel down and sift through the first two to three inches of topsoil where many interesting aircraft parts and personal items may be found. In the past, searches of the topsoil have revealed parachute "D" rings, seat belts, instruments, coins, shoes, Mae West life vests, ammunition shell brass, playing cards, canvas snaps, belt buckles, watches, wings and other insignia, and hundreds of miscellaneous items.

Souvenirs

Souvenirs may only be taken if the landowner permits it. If the wreck is on public land, always check with the local administrators as to their policy on aircraft part removal. Never use the term "artifact," as everybody will have you filling out forms and possibly denying you, and others, permission to visit the wreck.

When you travel to a crash site, you may be pretty pumped up. Going in, an hour's walk can seem like 20 minutes if the route is downhill, but coming out may be more difficult . If the landowner has given permission to remove souvenirs and it's a long walk out, take only what you can carry, preferably in a small backpack — you never know when you'll find a loose propeller that needs a home.

The wreck you find today may yield many parts that can be used as patterns in aircraft restorations. If you do find a rare aircraft, contact your local aviation museum and see if they need any parts, or if they know of anyone looking for parts.

Placing Memorials

Although the cost will have to come out of your pocket, placing a memorial is really worth the effort. You will feel a personal satisfaction to have researched a crash, visited the site, and payed tribute to those who lost their lives. Contact the landowner regarding a suitable location. Many public lands will allow a small plaque at the site plus a more suitable one, which may incorporate pieces of the aircraft, at the local headquarters.

Recovering Armament: Don't

During World War II, thousands of aircraft flew daily on stateside training missions. If an aircraft went down, the primary focus was to rescue the survivors, remove any of the crew who might have perished, and then secure the bombsight and avionics. Most World War II crashes were in pretty remote areas, and the military had no interest in lugging out the heavy armament. Many of the wrecks have yielded internal parts for machine guns and cannons, but never a whole one. If you do find one, be sure to have your picture taken with it.

The law is very specific about citizens owning machine guns, even old, rusted or bent ones. The Department of the Treasury's Bureau of Alcohol, Tobacco and Firearms (ATF) puts out an informative booklet entitled (Your Guide To) *Federal Firearms Regulations.* You can call your nearest ATF office — listed in the front of the phone book — to get a copy, and many gun shops also stock copies.

In a nutshell, the guide states that you may not possess or transport any machine gun that was not registered in accordance with the 1968 National Firearms Act. (Weapons, such as machine guns, are referred to as NFA firearms.) Basically, if your gun was on a hillside in 1968, it does not qualify for NFA status and must be donated to a museum or destroyed. Since it came from an aircraft crash, you may save it from destruction by proving its historical significance. Your best bet is to leave it where you found it and to contact the Bureau of Alcohol, Tobacco and Firearms. The bureau can arrange for you to show them where the weapon is located and donate it to a museum.

This 20mm cannon makes a tempting souvenir, but the jail time is not worth it. Photo by Gary Larkins

Although rarely, an unserviceable aircraft cannon or machine gun can be registered as a "Curio or Relic" through the Bureau of Alcohol, Tobacco and Firearms' Firearms Technology Branch, Room B230, Bureau of ATF, Washington, D.C. 20226. If this is the route you plan to take, photograph the weapon intended for recovery at the site. Photograph it from all sides to show that the weapon is indeed unserviceable. Leave the weapon at the site until you can obtain a ruling or are able to donate the weapon to an approved museum.

Remember, recovering, transporting and possessing a machine gun, its frame, or receiver is illegal. If convicted of possession of such a weapon, you may be subject to imprisonment and a $10,000 fine. Consult your local, state and federal laws before recovering any automatic weapons.

Photography At The Site

Take a camera with you, but avoid expensive, fragile or sensitive equipment. A 35mm point and shoot is fine, but a rugged 35mm SLR camera with a 28-85mm zoom lens is best. If your wreck is in the hills or in a forested area, I recommend using fast film, ASA 400 or more, plus a flash. This will light the wreck as well as fill in the faces of your companions.

Heading Home

No matter where the wreck is, you must always take home a souvenir. Inevitably, someone wants to bring back some choice piece, which usually weighs 200 pounds. Here's a good rule to follow when wreck chasing: "Only bring back what you can comfortably carry." Too many times people have tried to carry out something heavy, they get tired, and then they leave the souvenir miles from the wreck. This misplaced piece could be the rare part that is missing in the restoration of a potentially flyable aircraft.

After you have had your fill, gather everyone together and police the area. Do not leave any litter, paper, garbage or food wrappings. If you have done any digging, try to fill in the holes and brush the area with loose branches to cover your excavations. Restore the site to the way you found it to preserve the site for those who will follow and remember to turn painted surfaces down, away from the elements.

Removing The Entire Wreck

Not satisfied with just a few souvenirs? How about the entire airplane? If the wreck you have located looks like it could be rebuilt, you will need to acquire title to the aircraft, as well as the landowner's permission, before removing it. For a military wreck, sometimes a letter and set of photographs describing the condition of the aircraft will get the ball rolling towards a clear title. If the Air Force Museum, Wright-Patterson Air Force Base, Ohio, or the Naval Aviation History Museum, Pensacola, Fla., does not have an interest in the aircraft and it has not been deemed historically significant, a release can be issued in a few weeks. But, according to a Naval Air Systems Command public affairs officer, wreck chasers must obtain U.S. Navy permission before recovering an aircraft. The Navy reasons that "many of these aircraft went down with crew members on board and these aircraft are considered grave sites."

One of the first groups involved with the historical research of a Naval aircraft crash site is the Navy Bureau of Personnel's Office of Decedent Affairs, which will review records of the crash and the ultimate fate of the crew.

By law, the Navy never abandons equipment, including aircraft, without a formal act of abandonment, and the mere passage of time or lack of recovery effort by the Navy does not constitute legal abandonment.

The National Historical Preservation Act of 1966 is also a factor in exploring and salvaging aircraft. The law may require the Navy to preserve a crash site if the site is deemed of historical value.

As for civilian aircraft, you will need to contact the FAA to find out the name of the last registered owner. In the case of a crash, the title of the wreckage may have been transferred to an insurance company.

Civilian or military, your best bet is to consult an aviation attorney before removing any aircraft.

Deciding what to take with you depends on how much you are willing to carry. Before leaving the 6,500-foot crash site of B-17F, 42-5318, at 3955N 12306W, it was decided to only take what would fit in a jacket pocket or small backpack. The mode of transportation, here a Cessna 172, dictated only small, lightweight souvenirs could make their way home. From left, Taigh Ramey holding ball turret ring and waist .50-caliber machine gun mount, Nick Veronico, and Ed Davies. Photo by Murray Moore

McDonnell F2H-4 Banshee

Bureau Number 127614

Two views of F2H-4, Bureau Number 127614, at Marine Corps Air Station Kaneohe, Hawaii., in March 1956. Less than two years later, the aircraft crashed north of the town of Saratoga, Calif.

Photos by Clay Jansson

Researching a crash from start to finish

Following the instructions shown earlier, we will now go step-by-step and locate McDonnell F2H-4 Banshee, Bureau Number 127614. This aircraft was found on the wreck locator (see Appendix I, Aircraft Wrecks By State) at 3717N 12207W, crash date February 22, 1959.

Background On The F2H Banshee

The background is a deciding factor — if you have no interest in the Banshee or 1950s Naval Aviation, why go?

On March 22, 1945, the U.S. Navy contracted with McDonnell Aircraft Corp. of St. Louis, Mo., for two XF2D-1 prototype aircraft, a larger, improved version of the FH-1 Phantom. The first flight of an XF2D-1 Banshee was on January 11, 1947. The first Banshees became operational in March 1949 with VF-171. The F2H Banshee was affectionately known to its crews as the *Banjo*.

Exactly 150 F2H-4s were built, and the last was delivered on September 24, 1953. A total of 892 F2H Banshees of all types were constructed. The F2H-4 was the Navy's standard carrier-based, all-weather air superiority fighter until 1959. The last Banshees in the Naval Reserve were phased out in 1961 at NAS Oakland, Calif.

Specifications

Length: 48 feet 2 inches
Wingspan: 45 feet with tip tanks
Height: 14 feet 6 inches
Crew: one pilot
Armament: four 20mm cannons plus two hard points capable of carrying one 500-lb. bomb each
Weight (empty): 13,183 lbs.
Weight (max.): 28,500 lbs.
Max. Speed: 527 mph at sea level
Cruise Speed: 501 mph
Range: 1,716 miles
Service Ceiling: 46,500 feet
Engines: 2 x Westinghouse J34-WE-38 developing 3,600 static thrust ea.
F2H-4 Bureau Nos: 126351/126353, 127547/127693

Obtain The Newspaper Account

On a foggy February 22, 1959, Lt. j.g. James F. Wyley, 26, crashed F2H-4 Banshee 127614 in Stevens Creek, north of the town of Saratoga, Calif. The *San Mateo Times* reported that both wings were torn away as the aircraft passed through trees near Skyline Boulevard and that the wings were recovered about 100 yards east of Skyline. In reality, the left wing rests about 20 feet above the tail.

The one-wing fuselage carried Lt. Wyley about one mile from the Skyline impact point into Stevens Creek Canyon, where the aircraft struck the northeastern side.

Lt. Wyley was in a flight of two VAW-11 aircraft flying from NAS North Island, San Diego, Calif., to Alameda Naval Air Station. Ens. L. L. Marshall was Lt. Wyley's wingman on the flight. Newspaper accounts report that Ens. Marshall watched Lt. Wyley's aircraft descend into "the fog and rain." He immediately radioed a rough location of the missing aircraft to Moffett Field. Banshee 127614 and Lt. Wyley's body were found after a five-hour search by Santa Clara County Deputy Sheriffs.

Navy Probes Plane Crash

Navy officials today started an investigation into the crash of a jet fighter plane near fog-shrouded Skyline boulevard just south of the San Mateo county line in which the pilot was killed.

The twin-jet Banshee, piloted by Lt. (j.g.) James F. Wyley, 26, of San Diego, swooped low over the highway late yesterday and slashed a swath through high trees along the Skyline. Both wings were torn from the fuselage by the trees.

The airplane's wings were recovered near Skyline boulevard, one only 100 yards from the road.

The jet's speed rocketed the fuselage a mile further on, into wild Stevens Creek canyon north of Saratoga. The charred jet and the body of the pilot was found by Santa Clara county deputy sheriffs who searched for five hours after the plane crashed.

Naval officials said Wyley was in one of two jets flying from North Island Naval Air station in San Diego to Alameda Naval Air station. The pilot of the other plane, identified only as Ensign L. L. Marshall, said he saw the other plane go down in the fog and rain and immediately radioed Moffett Field Naval Air station.

Lt. Wyley is survived by a wife living in Coronado.

February 23, 1959, San Mateo Times *showing coverage of the previous day's crash.*
Ed Davies Collection

Obtain The USGS Map

Upon locating the coordinates on the USGS map, we found that the aircraft now rests in the Saratoga Gap Open Space Preserve administered by the Mid-Peninsula Regional Open Space District. A call to the district's office put us in touch with Ranger Patrick Congdon, who was very knowledgeable and familiar with the site.

The Aircraft History Cards For 127614

Card No. 1 — Contract Number: J1023 — Acceptance (Day Mo. Yr.): 11 06 53 — Model: F2H-4 — Serial Number: 127614 — STRICKEN

Effective 30 Apr '55 OpNav X Data For Col. "Action Code" Change To Include "Action Code" And "Sequence No." Also Include No. Of Tour Extensions Under "Service Tour" Heading.
Effective 1 Aug '55 ... Y Data For Fly. H... Qtr Will Show Under "Control" Heading; And Landings: 1—Arrested; 2—Water; And 3—Other Will Show Under Heading "Y": Hrs Flown Last Three Months; Cols. 1, 2, And 3.

Card	Reporting Custodian Name	Location	Cont. Cust. Code		Status	Action Code	Control	Y: Date of Inventory (Day Mo. Yr.)	Number of Tour	Y: Age in Months	Y: Accumulated Hrs. in Tour	Y: Accumulated Hrs. Since Acceptance	Y: Hours Flown Last Three Months 1	2	3	
H	[illegible] Mtd	St. Louis	88	4	B	A		11 06 3	0	063						127614
S1	[illegible] Mtd	St. Louis	88	4	C	C		15 06 3	0	063						127614
H	O&R BuAer Mtd	San Diego	88	4	B	R		16 06 3	0	063						127614
H	VMF 214	El Toro	29	4	A1	R		20 06 3	1	124						127614
P	VMF 214	EL TORO	29	4	A1			8 3	1	002	84	89	1	36	47	127614
P	VMF 214	EL TORO	29	4	A1			11 3	1	005	212	217	47	39	42	127614
N	VMF 214	EL TORO	29	4	AJ	C		02 02 4	1	35						127614
N	VMF 214	EL TORO	29	4	A1	C		24 02 4	1	34						127614
N P	VMF 214	KANEOHE / EL TORO	29	4	A1	C		01 03 4	1	457	286	290	50	14	9	127614
N	VMF 214	KANEOHE	29	4	A1	C		01 04 4	1	45						127614
N	VMF 214	KANEOHE	29	4	A1	C		24 08 4	1	075						127614
P	VMF 214	KANEOHE	29	4	A1			25	1	18	717	701	36	49	41	127614
33	FASRON 117	BARBERS PNT	21	4	EJ	F		06 06 5	1	75						127614
33	FASRON 117	BARBERS PNT	21	4	F6	F		17 06 5	1	75						127614
33	O&R BUAER M&S	SAN DIEGO	88	4	E1	F		28 06 5	1	65						127614
33	O&R BUAER M&S	SAN DIEGO	88	4	D1	X		08 07 5	1	65						127614
Y	O&R BUAER FA	SAN DIEGO	88	4	D1		2	31 03 5	1	21	828	847			6	127614
33	O&R BUAER FA	SAN DIEGO	88	4	B	KO		22 09 5	1	65						127614
33	FASRON 8	ALAMEDA	21	4	C9	CO		26 09 5	2	107						127614
X	VMF 214	KANEOHE	29	4	AJ	R		9 11 5	2	117						127614
X	VMF 214	KANEOHE	29	4	A1	X		7 12 5	2	107						127614
Y	VMF 214	KANEOHE	29	4	A1		154	29 02 6	2	3	156	11001			95	127614
	VMF 214	KANEOHE	29	4	AJ	X		27 7 6	2	117						127614
	VMFAW 214	KANEOHE	29	4	A1			26 02 7	2	14	621	1465	9		215	1276..
X	FASRON 117	BARBERS PNT	21	4	F6	R 0		11 10 7	20	11 7	0					127614
X	O&R BUAER FA	SAN DIEGO	88	4	E6	F 0		22 10 7	20	10 7	0					127614
X	O&R BUAER FA	SAN DIEGO	88	4	D1	X 0		26 11 7	20	10 7	0					127614

Aircraft History Card 1.

Card No. 2 — Contract Number: 51023 — Acceptance (Day Mo. Yr.): 11 06 53 — Model: F2H-4 — Serial Number: 127614 — STRICKEN

Card	Reporting Custodian Name	Location	Cont. Cust. Code		Status	Action Code	Control	Y: Date of Inventory (Day Mo. Yr.)	Number of Tour	Y: Age in Months	Y: Accumulated Hrs. in Tour	Y: Accumulated Hrs. Since Acceptance	Y: Hours Flown Last Three Months 1	2	3	
Y	O&R BUAER FA	SAN DIEG	8	4	D1		1	28 02 8	20	22	699	1943				127614
X	O&R BUAER FA	SAN DIEGO	88	4	DA	X 0		26 03 8	20	10 7	0					127614
X	O&R BUAER FA	SAN DIEGO	88	4	B	K 0		10 04 8	20	10 7	0					127614
X	VF 121	MIRAMAR	21	4	A1	G 0		14 04 8	30	09 9	0					127614
X	VF 121	MIRAMAR	21	4	A1	K 0		01 06 8	300	02	0					127614
X	VF 121	MIRAMAR	21	4	AJ	X		10 09 8	300	02 0	0					127614
X	VF 121	MIRAMAR	21	4	A1	X		22 09 8	300	02 0	0					127614
X	VAW 11	NORTH ISLAND	21	4	AJ	R		25 09 8	300	04 0	0					127614
X	VAW 11	NORTH ISLAND	21	4	A1	X		30 09 8	300	03 0	0					127614
Y	VF 121	MIRAMAR	2	4	A1		53	31 8 8	3	5	70	1617		35		127614
X	VAW 11	NORTH ISLAND	21	4	A1	K 0		1 11 8	300	02 0						127614
Y	VAW 11	NORTH IS	2	4	A1		6	30 11 8	3	7	76	1625		5		127614

STRIKE CODES 1AAC OPNAVINST 5442.2
DATE
TOTAL HRS FLOWN 1719
TOTAL MOS OPTG
STRICKEN FROM INVENTORY 2/59
OPNAV X 28 Feb 59 CUSTODIAN
ACTION 28 Feb 59 VAW-11 North Island
ACCIDENT OR RETIREMENT
COMBAT THEATER
REMARKS:

Aircraft History Card 2.

Shown are the date of acceptance, its squadron services and dates, as well as the date it was stricken and the total airframe hours.

Nick Veronico Collection

The Aircraft History Card

The information contained on the aircraft's history card, which was obtained from the U.S. Navy, shows that Bureau Number 127614 was constructed at McDonnell Aircraft Corp.'s St. Louis, Mo., assembly line. The aircraft's operational history is as follows:

11/06/53	Accepted by the U.S. Navy, St. Louis, Mo.
06/22/53	Delivered to VMF-214, MCAS El Toro, Calif.
03/01/54	Deployed to MCAS Kaneohe, Hawaii, with VMF-214
06/06/55	Assigned to FASRON-117, NAS Barbers Point, Hawaii
06/28/55	Flown to NAS North Island, San Diego, Calif., for overhaul
09/26/55	Assigned to FASRON-8, NAS Alameda, Calif.
12/07/52	Assigned to VMF-214, MCAS Kaneohe
10/11/57	Transferred to FASRON-117, NAS Barbers Point
10/22/57	Flown to NAS North Island, San Diego, for overhaul
04/14/58	Assigned to VF-121, NAS Miramar, Calif.
09/25/58	Assigned to VAW-11, NAS North Island, San Diego
02/22/59	Crashed, Fatal, Stevens Creek Canyon, Calif.
02/28/59	Stricken from Navy records with 1,719 hours total time

Key to Navy terms: VAW - Navy Airborne Early Warning Squadron; VF - Navy Fighter Squadron; VMF - Marine Corps Fighter Squadron; FASRON - Fleet Aircraft Service SquadRON.

A View Of The Crash Site

Left wing panel that was not "recovered by Skyline Boulevard" as reported in the San Mateo Times' *article.* Photo by Nick Veronico

Looking down on the final resting place of 127614 from the middle of the hill. The left wing panel is to the rear of the photographer, while the engines and canopy are at the bottom of the ravine. Photo by Ron Strong

Tail section of 127614 showing the Civil Air Patrol's orange "X" on the tail to show over-flying pilots that this crash has been located. This orange "X" is referred to as "marked" on many crash locator lists.

Photo by Nick Veronico

PART II

The fate of the Philippine Clipper

Philippine Clipper in flight over Alameda, Calif. Photo courtesy Pacific Aerial Surveys via Gordon Werne

When Pan American issued the specifications for the successor to the Sikorsky S-42 Flying Boat, the company was making aviation history. Pan American needed a larger aircraft to service its expanding Pacific routes — an airplane that could carry enough passengers, cargo, mail and fuel on the 2,400-mile journey to the Hawaiian Islands economically. The new design should also be able to fly the route with the style and luxury that travelers of the 1930s demanded. Pan American's concept heralded the dawn of a new era in aviation and opened up new corners of the globe to air travel.

Pan American's specifications resulted in the largest aircraft of its day, the Martin M-130. With a wingspan of 130 feet and a length of 90 feet 10 inches, the M-130 could sleep 18 to 30 passengers in berths. Passengers also enjoyed a lounge and air-conditioned passenger cabins. With a crew of seven men, the M-130 weighed almost 26 tons, carried 4,000 gallons of fuel, and her four Pratt & Whitney R-1830s propelled her through the skies at a cruising speed of 130 mph.

Pan American agreed to buy three aircraft from Martin at a reported cost of $430,000 each. The *China Clipper*, a name that would be used to describe both the Martin M-130 and the later Boeing B-314, was the first of the type to fly in December 1934, only seven years after Charles Lindbergh had soloed across the Atlantic. In November 1935, the *Philippine Clipper*, registered NC14715, was delivered to Pan American, followed three months later by the last of the Martin M-130s, the *Hawaii Clipper*.

Whenever a "Clipper" aircraft would leave for the Pacific, it was accompanied by much media fanfare. The *Philippine Clipper* began service on the San Francisco to Manila run and later pioneered the way to New Zealand and Hong Kong. She kept up the Pan American tradition of first class service, while leading the way across the Pacific as war clouds darkened in her path.

War Breaks Out, The End Is Near

The *Philippine Clipper* left Wake Island (located west of the International Date Line) on December 8, 1941, at 7 a.m. for a scheduled flight to Guam. Ten minutes later, Wake Island radio ordered her to return immediately. Pearl Harbor, the Philippines and Midway Island were under Japanese attack.

Because of the confusion created by these actions, no one at Wake was exactly sure what the next move for the

Philippine Clipper should be. At 11:30 a.m. Wake Island time, a flight of 36 Imperial Japanese Navy twin-engine Mitsubishi "Betty" bombers from Kwajalein flew in at wave-top altitude and attacked the island. The Wake Island Airfield and its U.S. Marine Corps Fighter Squadron suffered heavy damage from the Japanese attack. The *Philippine Clipper* took between 20 and 60 7.7mm machine gun hits during a strafing pass. Somehow, none of the bullets did critical damage, and the *Philippine Clipper* was still fit to fly. Mechanics stripped the aircraft's interior of all nonessential weight, and late that afternoon, the Clipper lifted off from the island's lagoon with a crew of eight, 22 Pan American employees, five passengers from the Guam flight, three Army officers, and one enlisted soldier. Without the aid of radio navigation, she flew through the night to Midway Island to take on fuel. At first light, she continued toward her home base at Pearl City, Hawaii.

Within months of the United States' official declaration of war with Japan, the U.S. Navy conscripted the Pan American flying boat fleet and their crews as part of the Naval Air Transport Service. This meant that the Pan American crews were officially active duty members of the U.S. Naval Reserve. During 1942, the Pan American crews were returned to civilian status. The operation of the Clippers, which remained Navy property, was then contracted back to Pan American.

The most sought-after of the Pan American Clippers were the Boeing B-314s. Their range and cargo capacity made them ideal for carrying V.I.P.s and high-priority cargo across both the Atlantic and the Pacific routes. The Martin M-130s, on the other hand, did not have the range or the capacity of the newer Boeing 314s and were relegated to ferry flights between Hawaii and California. The brilliant silver and orange prewar paint schemes of the *China Clipper* and the *Philippine Clipper* were quickly replaced with a crudely applied coat of non-specular blue/gray paint with large rectangular American Flags on the forward section of their fuselages. Their distinguishing marks were the civilian registration numbers, the name *China Clipper* on the bow of the *China Clipper*, and a wound stripe on the bow of the *Philippine Clipper* for having survived the attack on Wake Island. The only other Martin M-130, the *Hawaii Clipper*, crashed at sea on July 28, 1938, somewhere between Guam and Manila. No trace of that aircraft was ever found.

The *Philippine Clipper*'s military service during 1942 had been uncomplicated, and January 1943 had started out even less eventful. On January 20, the *Philippine Clipper* sat tied to her pier at Pearl City, Hawaii. For nearly a week, a series of arctic storms had been making the weather over the California coast very treacherous. The immediate forecast was so poor that the local Pan American operations manager had sent the crews back to their hotels with no realistic expectation of when the next flight to California might leave. Around noon that day, Rear Adm. Robert H. English directly ordered Pan American to make the *Philippine Clipper* ready for immediate departure, which required the Clipper's crew to be rounded up.

No documented reason can be found as to why Adm. English insisted on leaving on such short notice. Likewise, there is no explanation about his lack of concern over the weather in California. It has been speculated that the admiral had scheduled a very high-priority meeting at the Mare Island Naval Shipyard near Vallejo, Calif., a major U.S. submarine base, on Thursday morning, January 21, 1943. Supposedly, the urgency of this meeting was created because of Japanese advancements in anti-submarine warfare equipment. U.S. submarine losses were growing, and the Navy was desperate for some form of electronic countermeasures.

Adm. English was a submarine veteran. After serving in the North Atlantic in World War I, he rose through the peacetime Navy to become the commandant of the Pacific Fleet Submarine Force. The names and ranks of those accompanying him on this California trip also lend credence to the theory of how important his meeting was: submarine commanding officer, Capt. Robert Holmes, U.S.N.; submarine commanding officer, Cmdr. John J. Crane, U.S.N.; submarine commanding officer, Cmdr. William G. Meyers, U.S.N.; submarine executive officer, Lt. Cmdr. George R. Stone, U.S.N.; submarine gunnery officer, Lt. Cmdr. John O. Coll, U.S.N.; Naval science professor, Capt. Donald C. Godwin, U.S.N.R.; staff attache to COMPAC San Diego, Cmdr. Francis L. Black, U.S.N.; and Navy civil engineer, Capt. Robert E. Thomas, U.S.N.

Also on the passenger manifest was Navy nurse, Lt. j.g. Edna Morrow, U.S.N. Lt. Morrow was not on Adm. English's staff, but had been waiting for a space available seat to return to the continental U.S. because she was suffering from terminal cancer. In the ultimate of ironies, she knew that she was going home to die.

The Pan American flight crew consisted of Pan American Capt. Robert M. Elzey, age 36; first officer Orvan K. Judd, age 23; second officer and navigator John R. Maynard, age 28; third officer and pilot George H. Abel, age 26; first engineer Leo A. Mackota, age 32; senior second engineer John J. Eage; first radio officer Clarence P. Thompson; second radio officer George W. Angus; and steward John G. Hill, age 33.

In all, 19 souls left Pearl City at 5:30 p.m. PWT (Pacific War Time) January 20, 1943. The *Philippine Clipper*'s intended destination was San Francisco Bay. In her tanks was sufficient fuel for 24.2 hours of flying. Her estimated time of arrival in San Francisco was 10:18 a.m. PWT,

January 21, 1943. These calculations gave her a seven-hour flight reserve.

The Pan American flight watch office at Treasure Island received a radio-telegraph (Morse code) message from Flight V-1104 at 5:35 a.m. PWT, giving its arrival time as 6:35 a.m. It seems that they had picked up a significant tail wind and were arriving three-and-one-half hours early. Satisfactory radio-telephone (voice) communications were established at 6:16 a.m., and at 6:18 a.m. Flight V-1104 again revised their ETA to 7:10 a.m. The establishment of radio-telephone communications is significant as the type of transmitter used was limited to short ranges.

At 6:22 a.m., Treasure Island advised Flight V-1104 that the current weather conditions precluded a night landing and suggested San Diego as an alternate: "We feel that San Diego is the only possibility, unless it would be possible to hold until after daylight. Landing conditions are such you could not land before daylight."

Receiving no reply, Treasure Island radio asked again: "How do you feel about San Diego? Impossible to land here before daylight."

Flight V-1104 replied: "Roger ... Thank you." Confused by this answer, the Treasure Island radio operator then asked: "Will you give us what you might intend to do as quickly as you have decided? Standing by."

Based on the National Transportation Safety Board's transcript of the radio transmissions between the *Philippine Clipper* and Treasure Island, there never was a straight answer from the Clipper as to exactly what Capt. Elzey intended to do.

The most common weather problem encountered on San Francisco Bay was fog. During such periods, the Clippers had a secondary base at Clear Lake, a large fresh water lake located approximately 100 miles northeast of San Francisco and about 40 miles inland from the coast. But the problem that morning was much worse than local fog. The storm was of such a magnitude that the approach to Clear Lake, which was situated between two large mountains, was obscured. Clear Lake was strictly a VFR alternative.

The *Philippine Clipper* was a true flying boat, which meant that she was unable to operate from hard runways that may have been available just to the north of the storm, which was traveling in a southeast direction. San Diego was probably the only realistic alternative for the *Philippine Clipper*. One major consideration, however, was that the end result of their flying to San Diego would have meant either an unpleasant overnight train ride for the admiral and his staff or remaining overnight in San Diego, with the intention of flying back to San Francisco the next day. Even so, there was no guarantee that the weather would clear sufficiently for a flight to San Francisco on January 22 or even January 23.

Apparently, two other alternatives were considered, both supported by the flight profile and subsequent radio transmissions. The first alternative, originally suggested by Treasure Island radio, was to have the aircraft loiter in the San Francisco Bay Area and re-evaluate the landing conditions after daylight. The second alternative called for the pilot to fly out over the ocean, then slowly sink below the overcast in an effort to find the straits of the Golden Gate and gain entrance to San Francisco Bay. Which of the two they were trying to implement is a matter of speculation. Based on the fact that they were flying north in the opposite direction, it is probably safe to say they were not planning an immediate trip to San Diego.

The next radio communication came at 6:50 a.m., when Treasure Island notified Flight V-1104 that its position was approximately 43 nautical miles west-south-

One of the two remaining Pratt & Whitney R-1830 radial engines at the wreck site. Souvenir hunters have removed a number of cylinders from this engine.

Donna Struck stands next to the largest piece of the Philippine Clipper wreck, the partially covered tail section.

Photos by Nick Veronico

west of San Francisco. Flight V-1104 acknowledged. The next position report from the *Philippine Clipper* said that they were over the Farallon Islands, about 25 nautical miles due west of San Francisco. There is no known documentation of how Treasure Island or the *Philippine Clipper*'s navigator fixed their position in either case.

The *Philippine Clipper* radioed Treasure Island this message at 7:15 a.m.: "Been over the area, but now taking a course due west."

Treasure Island seemed surprised at their change of course and asked: "You are taking a course due west to sea. Is that correct?" The Clipper responded, "That's correct."

Treasure Island then asked about flying conditions. Flight V-1104 answered: "Flying conditions very poor, require a change of altitude to 7,000 feet."

The final transmission between Flight V-1104 and Treasure Island took place at 7:20 a.m. The *Philippine Clipper* radioed: "We are not interested in setting up on Oakland yet. We would like to have a fix, please, so would you arrange that? Go ahead."

No further transmissions between the *Philippine Clipper* and any other station transpired after that call.

Approximately nine minutes later, the *Philippine Clipper*, Flight V-1104, impacted the side of a mountain at the 2,600-foot level, flying due north, approximately 140 miles north of their last position fix.

The NTSB report explained the situation: "Because the westerly magnetic heading of 270 degrees, which when corrected for the variation of 18 degrees east, would result in a true heading of 288 degrees. This, together with the extremely strong south-southwest wind, estimated by the U.S. Weather Bureau to have been between 60 and 80 mph at an altitude of 3,000 to 7,000 feet, would make it quite evident that he was actually proceeding on a track closer to 35 degrees." The NTSB report also states that Capt. Elzey, who had accumulated nearly 5,000 hours of command flying time, was not only unaware of his position and heading, but also his altitude. In short, he was officially lost.

James "Bo" Hiatt holds the original memorial plaque at the top of the ridge above the wreck site. Three minutes before impact, the Philippine Clipper passed over Hiatt's childhood home as he waited for the school bus.

Photo by Nick Veronico

The final seconds of Flight V-1104 were witnessed by several residents around the small town of Boonville, Calif. At approximately 7:30 a.m., Mrs. Charles Wallach was surveying the rise of a creek near her home when the Clipper burst through the overcast. She was only able to observe the navigation lights for a few seconds and then listen to the howl of the engines at full throttle, which faded to a low rumble. Wallach was the local air raid warden, and she attempted to call the sheriff's office to report the incident. The storm, however, had brought down the telephone lines, and the circuit was dead. The telephone remained out of service until January 25.

The failure of the Clipper to respond to the radio must not have worried the personnel at Treasure Island, for the normal routine for receiving the flight progressed on schedule. Pan American sent the receiving launch to the north end of San Francisco Bay to wait, while the radio operator continued to send "blind" messages to the flight.

The Civil Aeronautics Administration and the Western Sea Frontier Command were notified of the loss of radio communications at 10 a.m. A subdued air search was initiated. At 6 p.m., Navy officials were finally satisfied that Flight V-1104 could no longer be airborne and initiated a full-scale search, which eventually stretched from Eureka to Monterey and 225 miles out to sea.

The first local newspaper accounts of the search appeared on Saturday, January 23. The coverage was brief and only noted that the Navy was looking for a "large transport seaplane with 19 persons aboard."

On Sunday, January 24, the press released the names and photographs of the missing crew and passengers. By January 25, everyone's nerves were growing thin. The search was being conducted based on the last radio transmissions, which led everyone to believe that the Clipper had been relatively close to San Francisco and heading out to sea. Thousands of man-hours were being expended, and nothing had been found, not even an oil slick.

The two top Pan American executives at Treasure Island, chief pilot Joe Barrows and Western Divisional Manager Jack Tilton, headed up a team of experienced pilots that included Ken N. Beer, the chief pilot in charge of training. They began to formulate a "what if" theory about the Clipper's loss. From years of flying the Alaskan bush for Pan American, Barrows knew the unpredictability of arctic storms, and from the numerous times he had run-ins with the Navy over operating procedures, he and Tilton believed that Capt. Elzey had probably been flying under a great amount of stress.

Their common assumption was that Adm. English knew full well what the weather on San Francisco Bay was like when they left Hawaii and that he had forced them to leave anyway. Since he had displayed such a disregard for the advice of the Pan American crew then, he may have had the same disregard for Capt. Elzey's advice when they arrived in San Francisco. The radio transmissions did seem to indicate that they had chosen not to loiter near Oakland until daylight. That being the case, they felt that Capt. Elzey probably chose the option to "scud run," to fly just below the overcast to get the admiral to his meeting.

Barrows, Tilton and Beer concluded that the navigator may not have realized that the hurricane force winds aloft, which had brought them into San Francisco three-and-one-half hours early, were now pushing them north and east over land. When Capt. Elzey was told that they were safely out at sea, they were actually some 20 miles inland.

From the crux of the available information, their theory seems correct. The *Philippine Clipper* had broken out of the overcast directly over the Wallach farm, and upon realizing their predicament, the pilot gave the aircraft full climb power. Flying in broken visibility, their northerly course took them into Bell Valley, a box canyon. It appears that the pilot even tried to turn with the terrain. Unfortunately, the mountain's slope at the top of the valley exceeded the Clipper's climb rate. The *Philippine Clipper* missed clearing the crest only by about 200 vertical feet, impacting the hill in a nose-down attitude and catching fire. The steady rains and soaked underbrush quickly extinguished any secondary fires.

Once Mrs. Wallach was able to get her phone call through to the proper authorities, the Navy dispatched an NAS Alameda-based Martin PBM Mariner. Unable to fly low enough to positively identify the wreckage, a Lockheed Electra 10A, flown by Lt. Cmdr. Henry J. Hollenbeck, inspected the crash site from the air, and just five minutes before sunset, he pinpointed the Clipper's remains. Ten days after its loss, the *Philippine Clipper* was found.

Early the next morning, a U.S. Army detachment moved in and formed an armed perimeter around the wreckage. The soldiers were under direct orders to shoot anyone caught inside their perimeter without proper credentials. Until Adm. English's briefcase could be located, the military was taking no chances of its falling into the wrong hands. Days later, after the military had pulled out, several boys located the briefcase in the branches of a tall oak tree.

The final flight of the *Philippine Clipper* officially ended in tragedy due to the actions of Capt. Elzey, specifically "not having fixed his position at the time of entering San Francisco Bay, or anytime thereafter." It is interesting to note that this judgment was based on a report that never documents whether the *Philippine Clipper* actually flew over San Francisco Bay. Considering that at 6:50 a.m. (PWT), Treasure Island fixed the *Philippine Clipper*'s position as 43 nautical miles west-southwest of San Francisco, then the Clipper's next report stated that they were 25 miles due west of San Francisco, and the fact that they crashed more than 100 miles north of San Francisco 35 minutes later seems to document sufficient probability that Flight V-1104 never did overfly San Francisco Bay.

Whether or not Adm. English played any part in Capt. Elzey's actions is purely speculation. The general feeling among the Pan American pilots was that Robert McNair Elzey was a very professional and capable pilot, and the flight profile he flew was totally out of character for him. Radio transmissions in the final hour were out of keeping with the crisp and to-the-point exchanges that were the hallmark expected of Clipper captains. At best, the reports are a compilation of open-ended conversations that implied confusion and indecision.

The Crash Site

"Sure, I remember. I was just a boy. We were living in Yorkville at the time, and I was waiting for the school bus. It was just before 7:30 a.m. It was raining, and my mother heard an airplane approaching and called me and my sister out to see it fly over. I could see the lights as it passed overhead. Airplanes were a pretty rare sight in those days," longtime Boonville resident James "Bo" Hiatt said, recalling the day of the *Philippine Clipper* crash.

Record setting pilot Elgin Long, left, surveys an oil cooler and propeller blade from the Philippine Clipper held by "Bo" Hiatt. Photo by Ed Davies

Hiatt had been to the crash site of the *Philippine Clipper* a number of times in his youth. He knew the landowner, and he made arrangements to take the bus to the crash site for a closer look.

Since the crash site is on private property, prior permission must be obtained before traveling to the site.

Standing at the top of the hill and looking around, the first thought to come to mind is, "If they had only been 200 feet higher and 200 feet to the right, they would have made it." If they could have cleared this canyon, they would have had ample room to climb up to a safe altitude. From here, the terrain slopes downward into the Ukiah Valley.

Even though the crash happened almost 50 years ago, there is an eerie feeling to be at the scene where 19 people lost their lives. The canyon is quiet, and the colors of summer — reds, greens and browns — make a beautiful picture. In the distance to the west, the blue Pacific Ocean can be seen framed by white clouds in the valleys below.

There are no paths, so a path had to be cut on our way down to the site. Hiatt and Gary Urdahl, a local logger, led the way down into the valley. The first 75 feet of the descent into the canyon is perfect deer or rattlesnake country with large rocks, waist-high brush, and weeds. No wreckage was spotted yet.

At about the 100-foot level, 8-to-10-foot-tall trees replaced the brush and walking became easier. Navigating down the ravine under the trees, Hiatt pointed out the point of impact. It is a depressed area with telltale traces of molten aluminum strewn in a 6-foot-diameter circle. There is not much to see at the point of impact, so we moved on down.

Another 30 feet down, the tail cone and horizontal stabilizer came into view. This large piece has what appears to be an inspection hatch or bulkhead large enough to walk through. Here, small pieces of corrugated skin and a set of cowling cooling doors are visible, sticking straight up as if a flag to mark the final resting place of one of the buried Pratt & Whitney R-1830s.

Descending another 30 feet around a small bend is a large area strewn with wreckage. Huge portions of the fuselage and the strut from the wing to the "water wing" are immediately recognizable. This area is massive, stretching a good 300 feet down the gully and 200 feet out into the underbrush. One of the R-1830s is visible even after having had a few cylinders cut from it over the years. About 10 feet down from the engine is the propeller spider and shaft, all three blades neatly hacksawed off by earlier souvenir hunters. We sifted through the large debris field for recognizable artifacts. A seat, a window frame from a pilot's window, an engine intake manifold, and many other easily recognizable aircraft parts were found.

Returning to the point of impact, we began to dig — not deep, mostly moving the top three to five inches of soil. At about two inches in depth, five feet out from the estimated point of impact, many snaps from a canvas cover came to the surface. Then a shout of amazement as an officer's coat button was found only three inches down.

Digging at the point of impact revealed a number of coins that were dated the same year, a crew member's Pan Am belt buckle, and numerous other personal effects.

Photo by Ed Davies

Then another coat button, and another. Then a plastic shirt button from just below the point of impact. This was followed about 15 minutes later by one of the most unusual finds, a 1938 Buffalo nickel. Shortly thereafter, a 1938 penny came to the surface. Many other small pieces of history were recovered, from more snaps to belts or cargo belt buckles, to broken pieces of dishes. We only removed a small number of items from the wreck site, since we were limited to what could be carried in our pockets.

Except for the area where the surface digging took place, the site remained the same. The next good rain erased our foot prints and hopefully will yield more artifacts from the *Philippine Clipper*.

Since our trip to the *Philippine Clipper*, Lowell "Ed" Wilcox has started a Philippine Clipper Museum at the FAA Flight Standards District Office at Ukiah Airport. The display is open during normal business hours and contains many unique artifacts, including the only remnants of the once proud Martin Clipper fleet. The other two M-130s were lost. The *Hawaii Clipper* disappeared without a trace over the Pacific in 1938, and the *China Clipper* crashed in Trinidad in 1945 with 23 lives lost.

Ed Davies, Enrico Masagli, Nick Veronico

The search for Pearl Harbor veterans in Hawaiian waters

During World War II, a number of aircraft crashed on and in the waters around the Hawaiian Islands. In an effort approved by the U.S. Park Service, in conjunction with the USS Arizona Memorial, the University of Hawaii, the Oregon Air Museum, and Erickson Air Crane from Central Point, Ore., the benefactor, a massive aircraft recovery project was begun.

A two-month search of the waters off Oahu led the Erickson search team to the wrecks of two Mitsubishi A6M Zero fighters, one Kate torpedo bomber, a Corsair, an SBD Dauntless, a B-24 Liberator, and a pair of F6F Hellcats, plus a number of other rare aircraft. Several World War II aircraft wrecks were also located on land, including an F6F, an F4U Corsair, and a P-47 Thunderbolt. A second Republic P-47 and a Douglas B-18 Bolo, both of which were originally discovered in the late 1970s, were also resurveyed for possible recovery.

Recovery efforts for the three veterans of the Japanese attack on Pearl Harbor began in August 1992. The Kate, which rests in 110 feet of water with its torpedo still attached, and the two Zeros, each in just over 100 feet of water, will be recovered pending final arrangements and displayed at the USS Arizona Memorial. National Park Service divers also may survey one of the Japanese Zero fighters and possibly recover an outer wing panel.

The University of Hawaii leased one of its two deep-submersibles to Erickson to aid the search team's recovery efforts. The three-man *Pisces V* is operated by the university's Hawaiian Undersea Research Lab. The *Pisces V* is towed to the dive area on a catamaran barge, which lowers itself under the surface to allow the mini-sub to float free. Divers attach air bags and harnesses to an aircraft, which then floats up to a depth of about 10 feet. When the aircraft is stabilized just under the surface, the submersible catamaran barge will be lowered beneath the aircraft. Then the platform will ascend to the surface, where it will be either towed to shore with the aircraft secured to the barge or lifted from the barge by helicopter and flown to a nearby pier.

The *Pisces V* also will be used in the recovery of the Douglas SBD Dauntless dive bomber, which lies in 680 feet of water off the Oahu coast. The B-24 Liberator was located in the Maui Channel, resting on its belly in 180 feet of water. All four engines are still attached, and although the fuselage is complete, the forward area took a tremendous beating during the aircraft's successful ditching. The tail turret is covered in black coral, and its guns point upwards towards the surface.

The Corsair was found intact in Maunaloa Bay, covered with marine growth at a depth of 105 feet. The tail wheel and hook are down, and the canopy is open as if the pilot ran out of gas or could not extend his main landing gear and elected to ditch the Navy fighter. This aircraft will not be recovered, as it provides a source of income for local dive shops and free-lance dive guides who take scuba enthusiasts to the fighter wreck.

Aside from the three aircraft planned for display at the USS Arizona Memorial, discussions are underway to display several of the other recovered aircraft permanently at other locations in the Hawaiian Islands.

Nick Veronico

Overview of the sunken Corsair showing the fuselage, engine and propeller, and right wing.

Cockpit section of an F4U Corsair in 110 feet of water.

Photos courtesy Erickson Air Crane

DEPARTMENT
A. F. Form No. 14
Revised May 14, 1942

WAR DEPARTMENT

U. S. ARMY AIR FORCES

PORTLAND ARMY AIR BASE, PORTLAND, OREGON.

REPORT OF AIRCRAFT ACCIDENT

Accident No. 45-9-1-2

(1) Place Ten miles east of Bullrun on Sandy River, Oregon (2) Date 1 September 1944 (3) Time 0745 PWT

Aircraft: (4) Type and model P-63-A-7 (5) A. F. No. 42-69137 (6) Station PAAB, Portland, Oregon.

Organization: (7) 4th Air Force (Command and Air Force) (8) 432nd AAF BU (Group) (9) (FTR-RTU) SE (Squadron)

PERSONNEL

Duty (10)	Name (Last name first) (11)	Rating (12)	Serial No. (13)	Rank (14)	Personnel Class (15)	Branch (16)	Air Force or Command (17)	Result to Personnel (18)	Use of Parachute (19)
P	DORAZIO, Earnest N.	P	O-773586	2nd Lt.	18	AUS	4th AF	Fatal	No

RECEIVED
SEP 14 1944

(20) DORAZIO (Last name) Earnest (First name) N. (Middle initial) (21) O-773586 (Serial number) (22) 2nd Lt. (Rank) (23) 18 (Personnel class) (24) Air Corps (Branch)

Assigned (25) 4th Air Force (Command and Air Force) (26) 432nd AAF BU (Group) (27) (FTR-RTU) SE (Squadron) (28) PAAB, Portland, Ore. (Station)

Attached for flying (29) 4th Air Force (Command and Air Force) (30) 432nd AAF BU (Group) (31) (FTR-RTU) SE (Squadron) (32) PAAB, Portland, Ore. (Station)

Original rating (33) P (Rating) (34) 15 July 44 (Date) Present rating (35) P (Rating) (36) 7/15/44 (Date) Instrument rating (37) ——— (Date)

First Pilot Hours:
(at the time of this accident)

(38) This type	44:25	(42) Instrument time last 6 months	32:10
(39) This model	36:10	(43) Instrument time last 30 days	3:35
(40) Last 90 days	44:20	(44) Night time last 6 months	
(41) Total	282:05	(45) Night time last 30 days	

AIRCRAFT DAMAGE

Damage		(49) List of Damaged Parts
(46) Aircraft	WS	Acft. Burned
(47) Engine(s)	WS	" "
(48) Propeller(s)	WS	" "

(50) Weather at the time of accident 1800 scattered, 3500 overcast, visibility unlimited.

(51) Was the pilot flying on instruments at the time of accident No

(52) Cleared from P.A.A.B. (53) To Local (54) Kind of clearance CFR

(55) Pilot's mission Interception, Training mission.

(56) Nature of accident Spun in.

Page One of the Army Air Force's Aircraft Accident Report on P-63A, 42-69137.

Courtesy Sig Unander

P-63 Kingcobra

U.S. Army Air Force's Bell P-63 Kingcobra in flight. Courtesy Sig Unander

On September 1, 1944, the end of the war was in sight. The once mighty Japanese military machine was being forced back to its home islands, and the Northwest Pacific had become an armed camp during four years of war.

On the Portland, Ore., Army Air Base, the sky that autumn day was overcast, but visibility was listed as "unlimited." A flight of the 432nd Bomb Wing, Fighter-Reconnaissance Training Unit, was taking off on a training mission, an interception of a fictitious enemy aircraft.

Five Bell P-63 Kingcobras snarled down the runway and lifted off into the gray dawn at 7:10. Sleek, streamlined and a pleasure to fly, these fighters did not have enough high-altitude maneuverability to challenge the Mitsubishi Zero, but the rugged Kingcobra was ideally suited for stateside training of young fighter pilots.

Leading the formation of five as they flew off to the southwest was Lt. L.A. Chiles. His wingman was 2nd Lt. Ernest N. Dorazio of Blairsville, Penn., a student pilot with almost 300 hours of flying time, including 44 hours in the P-63. His Kingcobra, serial number 42-69137, had 320 hours flight time. It was getting a little war weary, the wings slightly buckled from the rough handling by the dozens of student pilots.

Making the "interception" at 10,000 feet, the formation turned east of the Bull Run Reservoir on the Sandy River as Lt. Chiles called the Portland Army Air Base for a vector home. The flight started a gradual descent into broken clouds at 8,500 feet.

"Some clouds have rocks in them," Lt. Dorazio's instructors had told him repeatedly. "In clouds, don't trust your senses — trust the instruments." Solid words of advice. Dorazio had a little more than 30 hours of instrument flight time.

The cloud layer was getting thicker. Looking for a way through, the flight descended into what looked, in the early morning dimness, like a large hole in the cloud. It wasn't. It was a thicker cloud. Instantly, the pilots lost visual contact with each other, and the formation broke up in the mist. Several tense minutes passed as they descended, each man alone in the swirling grayness, earth and sky invisible, trying to "trust the instruments."

One plane came out of the clouds in a steep dive. As the pilot saw a ridge rushing up at him, he yanked back on the controls, and his aircraft swooped back up into the mist, leaving him with the memory of that terrifying sight and distracting him from the instruments for a crucial instant. Was he climbing? Was he turning? In the gray disorienting void, his senses did not tell him.

When he came out of the clouds again, he had one wing down in a spin. As his flying mates watched, 2nd Lt. Ernest Dorazio's Kingcobra struck the ridge in a ball of fire.

At least Lt. Dorazio's family could take his body home for burial. His Kingcobra was bluntly written off as "aircraft burned." There was a war to finish, and the wreck was back in the hills, surrounded by second-growth timber, far from any road, and impractical to salvage.

Greater events erased the memories of one student pilot's death. As the war receded into the past, Kingcobra 42-69137 lay forgotten on its timbered ridge, and the young trees grew up around the scars of the crash. The few P-63s that survived the war were scrapped and melted down to make aluminum pots and pans.

Logger Kenny Valentine saw 69137 around 1981. It was still identifiable. The young trees that had sprouted around the crash site were now 40 years old, and the seedlings had become valuable timber.

The Mount Hood National Forest, having jurisdiction over the Bull Run Watershed where the P-63 lay, had arranged logging sales in the area.

As the area around the Kingcobra was logged, fallen trees were dragged back and forth over the wreck. It was reduced to scraps, and fragments of torn aluminum were scattered widely over the stump-speckled hillside. The logging boss carried away the fighter's Allison engine. One of the .50-caliber wing guns was removed by a young logger, taken home, and then tossed into his backyard, where it lay around for a few years before being hauled to the dump. Other loggers had picked through the remains looking for anything of interest to them, and some had cut off bits for souvenirs.

After the final load of logs was hauled away, smashed electronic boxes, landing gear, tatters of aluminum skin, and hundreds of tiny parts and fragments buried in the rutted soil or tangled in the blackberries were all that remained, though vague rumors of an old airplane crash somewhere in the Bull Run Watershed still circulated.

By 1988, Sig Unander had heard those rumors. Unander, a native of Cornelius, Ore., was born 30 years too late to have had a chance to fly the warbirds that fascinate him. He carefully collects newspaper articles about old air crashes, listens to the rumors, then he studies and cross-references the stories. If one sounds like it might pay off, he begins by interviewing elderly outdoorsmen and witnesses. Then he goes out into the rugged back country in an attempt to discover those few wrecks that have not been destroyed by military authorities or vandals. His purpose is to pinpoint the sites for historical purposes, to memorialize the now forgotten aircrews, and to salvage usable parts needed in the restorations of the few surviving World War II aircraft.

The rumors of the Bull Run wreck sounded interesting. In several expeditions to the watershed, he spent a lot of time slogging around on steep, brushy hillsides without finding any trace of an aircraft. Unander spoke to Bruce Rigby, a timber sales officer of the Columbia Gorge District of the Mount Hood National Forest, in the spring of 1989, and Rigby's curiosity was aroused.

Rigby, a longtime World War II aircraft enthusiast, had himself found a P-38 Lightning wreck in the watershed years before. He began asking around among the old loggers and other woodsmen he knew. Valentine told Rigby that he had seen an old wreck on Sandy Ridge when he helped log the site in 1981. Valentine took Rigby to the site, and Rigby then guided Unander to the shredded wreck.

When he first saw it, Valentine had been told it was a Bell P-39 Airacobra, the predecessor to the P-63 Kingcobra. But now this scattering of aluminum debris gave no hint as to its origin.

As Unander and two friends began to work the site, Rigby and Valentine stopped by to check on their progress, then stayed to help dig. There is something fascinating about old warplane crash sites — an evocation of the grim days of the war, the ever-present hope that a usable part might be found, and the recreation of a doomed aircraft's last moments.

Unander and his friends slowly explored the wreck site with shovels and a metal detector, recovering both burned rubbish and shiny, new-looking components still bearing their original paint and military part numbers. One of those numbers prompted Unander, on a hunch, to telephone Dan Chvatal, an aircraft historian and collector in the Midwest. Chvatal, the owner-restorer of two P-63s, consulted his old technical manuals and confirmed Unander's hunch. This was a P-63 Kingcobra, the successor of the P-39.

Now the mystery of how the aircraft got to Sandy Ridge and who was at the controls remained to be solved. Scanning the wartime issues of *The Sandy Post*, a local newspaper, Unander came across a front page story that ran on September 7, 1944, describing the fatal crash of a Lt. Dorazio. This information was passed on to the Air Force. About two months later, the Air Force sent an official copy of the War Department's crash report that described Dorazio's fatal crash. The aircraft serial number on the crash report matched the tail number found on the wreck.

The spot where the Kingcobra had struck was marked with a wing spar, a twisted propeller blade, and a surprisingly small patch of burned forest duff. The Kingcobra wreck yielded one main landing gear assembly and many airframe parts that will be used in the restoration of one of Chvatal's P-63s. The remaining scattered wreckage was simply the bones of a once-beautiful aircraft, a memorial marking some long-lost history.

Lt. Dorazio gave his life for his country as surely as if he had met a German or Japanese bullet. Yet, because he never went to war, he and his aircraft were forgotten until their story was brought back to light.

During the war years, student pilot errors, treacherous weather, or accidents brought down more than 300 military aircraft in the Pacific Northwest. Many have never been found. They lie hidden under the dense canopy of evergreen forests, entombed in ice on mountain peaks, and half-buried on the sea floor or on the silty bottoms of lakes and rivers. Occasionally one is discovered, and a few of its parts may go to a restoration of one of the few flying World War II aircraft that survive today. Other aircraft are out there, just waiting to be found.

Walt Wentz

Sig Unander at the crash site of P-63A, 42-69137, in the Bull Run Watershed, 30 miles east of Portland, Ore. Wreckage in foreground is vertical stabilizer and part of the aircraft's tail. Courtesy Walt Wentz

After visiting the wreck site, searchers Grantham and Register placed this memorial plaque to commemorate the wartime training loss and to explain the large chunks of metal to hikers who stumble onto the site.

Memorial for a Corsair pilot

On March 15, 1944, Goodyear FG-1A, Bureau Number 13380, sat on the ramp at the Simmons Knot Airport, N.C. The Pratt & Whitney R-2800-8 radial engine was rumbling over with its customary confidence, while in the cockpit, Marine 2nd Lt. Edward J. Riordan was completing his cockpit checks before a night familiarization training flight.

Riordan had accumulated a total of 355.3 flying hours, 48 of those in the Corsair. Bureau Number 13380 had 128.5 hours on the airframe, while the engine had 143.8 hours running time. The plane was assigned to VMF-324, Marine Air Group 34, 3rd Marine Air Wing. Unknown to anyone on the Simmons Knot ramp, disaster was soon to strike.

Corp. Donald S. Keefer was the plane captain for 13380 (which carried the training code 182-A on the fuselage) and on March 16, 1944, he recounted the events of the previous evening: "Plane Fox 182-A, of which I was plane captain, was scheduled for a night familiarization flight at 2200. On or about 2130, I made an inspection of the plane and found it satisfactory for flight.

"I was warming the plane up when Lt. Riordan came out to the plane. He also made an inspection of the plane and accepted it for flight. At his signal, I removed the chocks and guided him out to the taxi strip near Runway 22. Lt. Riordan moved out on the runway and prepared for takeoff.

"During takeoff, the engine of the plane was running smoothly, and as the plane was taking off, I was watching the tail light, which soon disappeared from sight. I started to turn and to continue on with my work when, suddenly, a burst of flames seemed to appear from the top of the trees. Upon investigation, this proved to be plane Fox 182-A piloted by Lt. Riordan.

"Plane Fox 182-A was flown by Lt. Riordan on the previous flight, and he did not note any defects that should be remedied before the next hop."

Another witness to the accident was 2nd Lt. William E. Lawhon who stated, "I saw Lt. Riordan's plane take off normally and watched his plane clear the trees at the end of the runway. Then I took my eyes off the ship for an instant, and as I looked for it again, I observed a tremendous explosion a short distance beyond where I had last seen his lights."

The destruction of the Corsair was investigated by the Marines. However, due to the pressure of war and training requirements, the pilot's body was removed along with a few salvageable bits and pieces, a report was made, and the incident was forgotten. Forgotten, that is, until the author and some friends rediscovered the wreckage of the FG-1A in the Croatan National Forest near New Bern, N.C., during August 1987.

Left to right, David Comer, Regan Register and A. Kevin Grantham display some of the artifacts recovered from the crash of FG-1A, Bu. No. 13380.

The remains of the aircraft had been basically untouched since the accident investigation and comprised a considerable amount of wreckage. Since that time, the author and Regan Register have coordinated two visits to this site with the Forestry Service in order to document as much of the plane as possible. They have also documented the wreck with a plaque for anyone who might discover the wreck in the future.

A. Kevin Grantham

The remains of the FG-1A's mighty R-2800-8 radial engine.

Note how casting was sheared during impact.

FG-1A Crash Site

Photos by A. Kevin Grantham

The center section of the wrecked Corsair could yield a number of parts that might serve as patterns in an FG-1A restoration.

The manufacturer's data plate was found among the wreckage.

U.S. Navy locates historic airship

The U.S. Navy located and photographed the wreckage of the airship *USS Macon* and two of her F9C-2 Sparrowhawk fighters in 1,500 feet of water north of Point Sur, Calif., on June 24, 1990. The Navy was aided by historical information and ocean current calculations provided by the Naval Postgraduate School and the Monterey Bay Aquarium Research Institute in locating the *Macon*. Searchers used the world's deepest diving manned submersible, the *Seacliff* (DSV-4), in their efforts. The *Seacliff* is a 32-foot-long, titanium-hulled, three-man vehicle capable of depths up to 20,000 feet.

The *Macon* was returning from maneuvers with the Pacific Fleet on February 12, 1935, when she encountered a squall. The turbulence broke the airship's vertical fin, tearing the spine of the airship and deflating the craft's helium cells, causing the airship to lose control and eventually crash into the sea southeast of Monterey. Only two of the crewmen aboard the *Macon* at the time of the crash were killed. The crash of the *Macon* ended the "Zepplin era" for the United States and its military forces.

The *USS Macon*, Navy designation ZRS5, was commissioned on June 23, 1933. She was powered by eight 560-horsepower Maybach engines with a maximum speed of 84 mph and a range of 9,200 miles at a cruising speed of 60 mph. The *Macon* carried a crew of 60 and was 785 feet long, 132.9 feet in diameter, and held 6,500,000 cubic feet of helium.

The USS Macon backs away from the mooring mast. Note the position of the engines. The front pair of engines is pointed towards the ground, while the other six engines are reversed.
Photo courtesy U.S. Navy

The *Macon* was the sister ship to the *USS Akron* - ZRS4, which was lost during night exercises in the Atlantic on April 3/4, 1933. Both airships were designed as flying aircraft carriers to scout ahead of the fleet and were equipped with four F9C-2 Sparrowhawk biplane fighters for defense.

The Sparrowhawks were lowered into the air by a trapeze and then launched for defense of the mothership. When a mission was completed, the Sparrowhawks would return to the airship and, by means of a large hook mounted to the upper wing, were recovered by flying the hook onto the trapeze. Once tethered to the airship, the aircraft were hoisted into the 70-by-58-foot hangar deck.

Since the airship's crash into the cold Pacific Ocean, there have been a number of attempts to search for the wreckage, all unsuccessful. As recently as last summer, an unsuccessful, yet fruitful expedition had been mounted by Hewlett-Packard's David Packard. Although Packard's expedition did not find the *Macon*, they did locate a fisherman, Dave Canepa, who had netted a "piece of aircraft wreckage" during the 1970s. Canepa gave the location where he had netted the piece of metal to the Monterey Bay Aquarium Research Institute, which in turn contacted the Naval Postgraduate School.

Using this information as a starting point, the *Seacliff* was able to locate, photograph and recover small pieces of the airship's framework for study.

The *Seacliff* and her mothership, the *M/V Laney Chouest,* went to Monterey to survey the underwater damage and to relocate submerged oceanographic instruments that were moved during the October 17, 1989, Loma Prieta earthquake. According to the Naval Postgraduate School, some of the instruments that were anchored in the undersea Monterey Canyon moved more than 600 feet.

The F9C-2 Sparrowhawk fighters are very significant, as only one example is known to survive, which is a composite aircraft currently on display in the Smithsonian. After the Navy's success in recovering the Grumman F3F-2 off the coast of San Diego, enthusiasts are hopeful that the Navy will attempt to raise at least one of the *Macon*'s Sparrowhawks.

Nick Veronico

The attraction of finding the USS Macon's final resting place is the hope that one or more of the F9C-2 Sparrowhawk fighters will be recovered.

Photos courtesy U.S. Navy

One of the USS Macon's four Sparrowhawk fighters perched on the "trapeze" that raised and lowered the fighters from the hangar deck. Note the shadow of the USS Macon above the aircraft's left wing.

Rare Grumman fighter recovered from the Pacific

Fifty years after ditching in the Pacific Ocean, F3F-2, serial number 0976, rests on her landing gear at NAS North Island, San Diego. The Grumman biplane is currently under restoration at the San Diego Aerospace Museum.

Photo courtesy U.S. Navy

On Tuesday, August 29, 1940, a young U.S. Marine Corps pilot attached to VMF-2 ran short of fuel just 200 yards from the deck of the *USS Saratoga* (CV-3) and ditched his Grumman F3F-2 biplane, serial number 0976, into the waters of the Pacific Ocean.

During the first week of April 1990, that same aircraft, with the pilot watching, was raised from the ocean off Del Mar, Calif. Recovery of this aircraft was the first step in a restoration process, with a goal to place the one-of-a-kind biplane fighter on static display at the Museum of Naval Aviation in Pensacola, Fla.

The aircraft was discovered on a routine training exercise and was recovered in conjunction with a salvage effort to retrieve a UH-1N helicopter, which ditched in October 1989.

The Marine Corps Grumman F3F-2 was one of only 82 built (including the first "X" model) and first flew in March 1935 as the improved version of the F2F, having a more powerful 750-horsepower Wright Cyclone engine. Although the F3F-2 served only in small numbers, it saw prewar duty in Marine Corps units on both coasts and at various Naval air stations. The Marine Corps received most of the 81 production aircraft, though the Navy did equip VF-6 with the F3F-2. Most of the F3F-2s were withdrawn from front-line service in May 1941 and were replaced with the Grumman F4F Wildcat. There was, however, at least one F3F-2 in Hawaii on December 7, 1941, but its fate is unknown.

The discovery and recovery of the F3F-2 has left some aviation historians "very excited," according to Maj. John Elliot, USMC, ret., head of the Aviation Branch, Naval Historical Center at the Washington Navy Yard. When originally discovered, the aircraft was viewed through a remote underwater television camera at a depth of more than 2,000 feet. The aircraft was remarkably well preserved, with the original paint and markings clearly visible.

At that time, the aircraft was almost completely intact.

However, sometime between the time it was discovered and recovered, some significant damage occurred, probably due to the aircraft's being caught and dragged by a fishing net. In the process, the plane's right wings, the canopy, and the tail were either torn off or damaged.

Robert Rasmussen, director of the Naval Aviation Museum, said, "It was very, extremely disappointing to have it on the bottom for 49 years in absolutely perfect condition and then come back to find that it's been torn apart by a fishing net."

The aircraft originally was delivered to the Marine Corps on December 21, 1937, at the Marine Corps Air Station Quantico, Va., and spent its entire service life on duty with the Marine Corps. The aircraft was so well preserved that the yellow stripe on the fuselage, indicating the sixth section's leader, was still bright, as were the squadron markings, 2-MF-16, and the national and Marine Corps insignia.

When 1st Lt. Robert E. Galer, then 27, experienced fuel problems and was forced to ditch, the aircraft had flown some 836.2 hours. Galer, now 77 and living in Dallas, Texas, was on hand for the recovery and received a pencil that was found in the cockpit. The former pilot was quoted as saying of the accident, "I just wanted to get down, then worry about being sorry. I was on the downwind leg when I switched on the reserve tank, and the engine just quit. There is not much you can do at 1,000 feet but go down.

"There were destroyers following the aircraft carrier on plane guard duty, and the landing speed was only 60 mph. The aircraft landed, nosed over, and then settled back. It floated for a while, and I got picked up," Galer said.

Galer recovered from the mishap and went on to receive the Medal of Honor for his destruction of 11.5 Japanese aircraft while flying a Grumman F4F Wildcat over Guadalcanal in October 1942. He ended the war with 13.5 enemy aircraft to his credit. He later flew jet fighters during the Korean War and retired from the U.S. Marine Corps in 1957 as a brigadier general and director of the Navy's guided missile program.

The stubby little Grumman was one of the Navy's last biplane fighters. Streamlined with an enclosed cockpit and retractable landing gear, the F3F-2 epitomized both the peak of biplane design and the twilight of the biplane era. The F3F-2 was proudly displayed by the Navy at the 1938 Cleveland Air Races.

The aircraft is now being restored by volunteers at the San Diego Aerospace Museum. The Navy will retrieve the aircraft in 1995 for permanent display in Pensacola.

The only other aircraft of this type is a Grumman G.22 Gulfhawk II, which spent its life with Gulf Oil as an aerobatics aircraft. The G.22 Gulfhawk II is a hybrid aircraft with the fuselage of an F3F-2, the wings of an F3F-1, and a 1,000-horsepower Wright R-1820 engine. The plane is now on display at the National Air and Space Museum, Washington, D.C.

Thomas Wm. McGarry

This Grumman F3F-2, serial number 0972, was photographed at Oakland, Calif., in April 1938. She served in the same squadron as the F3F-2 that was recovered off the coast of Del Mar, Calif., in April 1990. Photo by William T. Larkins

TF-51H returns from Arizona hillside

Lt. Robert G. Westlund had no way of knowing that he was making his final flight as he warmed up the Merlin in a TF-51H Mustang, serial number 44-64697, on a cold morning in December 1951.

Lt. Robert G. Westlund in the cockpit of a TF-51H, the same model Mustang in which he perished on the Arizona hillside.
Robert A. Kropp Collection

Westlund took off on a cross-country navigation flight from Fresno, Calif., to his duty station, the 3625th Air Force Central Training Squadron at Tyndall Air Force Base, Fla. Several hours into his flight, military officials in Arizona picked up a final radio transmission from him. He reported an oxygen system failure at high altitude and requested clearance for an emergency landing at Davis Monthan Air Force Base in Tucson.

A rancher told military authorities he heard a shriek and then an explosion that sounded like a bomb when the airplane impacted with the earth. Several weeks later, the pilot of a small plane reported sighting the wreckage of the fighter deep in a canyon on a remote mountainside in southern Arizona. Air Force officials sent an accident investigation team into the area, recovered the remains of the pilot, and photographed the wreckage. Within a month, the incident was entered into the military files and written off as an unfortunate and one of many all too frequent mishaps.

Lt. Robert G. Westlund

Arizona resident Richard Johnson, 38, a budding aviation enthusiast and military historian, was intrigued by the story of this missing Mustang, and he was determined to find the airplane. In June 1991, Johnson not only located the rare Mustang, but spent several days with a group of men removing the warbird from its remote crash site.

Johnson said he made the recovery with the hope that the vintage fighter would return to the air again one day. "Whether it can be restored or used as a parts source for another P-51H restoration project, the end result justifies this endeavor. Mustangs were meant to fly, and I think Lt. Westlund would have wanted it this way," he said.

The deceased pilot's family members agreed and traveled to Scottsdale, Ariz., from California to view the remains of Lt. Westlund's Mustang. "It was a very moving experience," Johnson said. "I had been in contact with Lt. Westlund's family about six months prior to the discovery of the airplane. I had spoken with them to tell them about my intention of placing a memorial plaque in a town not far from the crash site in the mountains, and they were pleased that I cared about something that had happened such a long time ago.

"They were kind enough to send me copies of the military accident report, and the pieces of this puzzle started to come together. I searched for the Mustang on several occasions before I actually found it, and after coming upon something like this that was the scene of a tragedy 40 years ago, I had my reservations about trying to remove the airplane. I came to the conclusion that a person can still show reverence to the

Crash Site Of The TF-51H Mustang

The wreckage of 44-64697 yielded two rebuildable wings, the tail section, and numerous cockpit and fuselage parts. The Mustang will undergo extensive restoration in Texas. Robert A. Kropp Collection

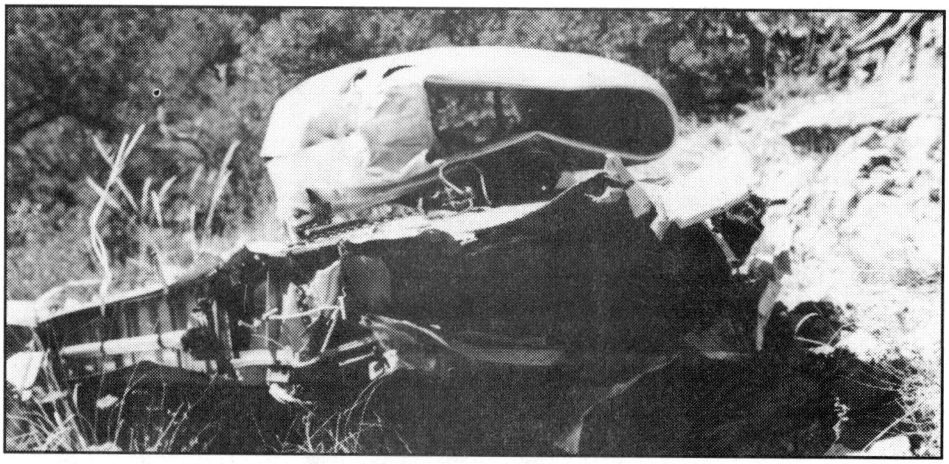

The Mustang was transported to a covered hangar in Scottsdale, Ariz., where Johnson and his partners examined and inventoried their find. Original ammunition boxes were still in place and undamaged, although the original armament group of .50-caliber Browning machine guns had been removed before the final flight.

Robert A. Kropp Collection

North American TF-51H Mustangs lined up on the ramp at Tyndall Air Force Base, Fla. Lt. Westlund's AAF SN 44-64697 is the first from left. Robert A. Kropp Collection

misfortune of others and help preserve a little piece of aviation heritage at the same time."

The P-51 Mustang, developed during World War II for its role as bomber escort, is perhaps one of the better-known fighter planes of the war. The first Mustangs arrived in England in December 1943, and after a series of production line refinements, the fighters were instrumental in turning the tide of Luftwaffe air superiority. Both the Army Air Forces and the Royal Air Force used the Mustang in a variety of combat roles.

The early variants were powered by 12-cylinder, in-line Allison engines; the later models by the Rolls-Royce pattern Packard-Merlin. The final production P-51, designated as an "H" model, was the Cadillac of Mustangs. A refinement of the P-51D, the "H" was a completely new and different airplane, and only the engine spacers of the earlier variant were interchangeable.

A sleek and fast-flying gun platform, the "H" model is longer, lighter, and took command of the sky with a 1,490-horsepower V-1650-9 Packard-Merlin, with water injection and improved cooling for extended operation at war emergency power settings. Developed late in the war, 555 "H" model Mustangs were built and delivered to the Army Air Forces by November 1945. North American Aviation had a standing contract for an additional 1,445 "H" model Mustangs, but the agreement was canceled in September 1945 with the surrender of Japan.

However, the Mustang saw combat again in the Korean War, and while the earlier models were pressed into service under a fighter-bomber role, the "H" models remained stateside and were used as trainers. The Air Force redesignated the P-51 as F-51 — the "F" standing for "Fighter," whereas the "P" meant "Pursuit." The "H" model's light weight was seen as a drawback in a ground attack role, and the Mustang continued in service with Air National Guard units before being phased out completely in 1957. Lt. Westlund's Mustang was listed on the government rosters as a "TF-51," the "T" prefix indicating the airplane was a training ship and was not armed.

During the mid-to-late 1950s, most of the "H" model Mustangs succumbed to the scrapper's torch. Of the 555 "H" models built, only six are known to remain, and only two of that half-dozen are in flyable condition.

Johnson at first searched on foot unsuccessfully for the Mustang, which was located on federal land. "I had a friend call me and tell me that he had found the wreck," Johnson said. "He had gone into the area and had found some small pieces of metal, but nothing like what I saw in the fuzzy pictures in the military accident report. It just didn't jive."

An R22 Robinson helicopter was chartered for the reconnaissance mission, and after four hours over the crash site, the Mustang was spotted on a warm June morning. "It was really like looking for a needle in a haystack," Johnson said. "There's an awful lot of mountain, scrub oak and chaparral bushes out there that can hide an airplane. You can literally be within several feet of something and not see it. When I saw a little tiny piece of aluminum wing root glinting in the sun, I knew we had found what we were looking for."

The accident report indicated the Mustang had broken

Aviation historian Richard Johnson standing near the vertical stabilizer of the TF-51H Mustang he helped to recover. Despite almost 40 years of exposure under the Arizona sun, unit markings are still discernable. Robert A. Kropp Collection

apart in a supersonic dive, a fact confirmed by Johnson and his two partners, Neil Jones, chief helicopter pilot for Arizona Wing and Rotor, and Scott Urschel of Urschel Manufacturing Inc. The Mustang's wings were hidden in underbrush several miles apart from one another, and the rest of the wreckage was scattered over several miles.

The trio then chartered a trip to the wreckage and spent two days photographing and cataloging their find. Although the Mustang had suffered extensive damage in the crash, some items were found in relatively good condition. "I had never been to an airplane wreck that hadn't been picked through by souvenir hunters before," Johnson said. "We found everything just as the military had left it; the throttle was cracked wide open, the radios were still tuned to their final frequencies, and the Merlin crankcase contained gummy oil. The armor windscreen was up against a bush, and the pilot's seat armor was resting against a rock. Access panels could still be opened with a Dzus tool. The feeling of elation in finding this wreckage undisturbed was sombered by the discovery of Lt. Westlund's parachute harness and the remains of his parachute. It put things into perspective pretty darn quick."

Johnson and his two partners decided the Mustang was worth saving and then went through a six-week process to obtain the airplane legally. "It was a real paper chase," Johnson said. "Because the airplane was located on wilderness land, I first approached the United States Forest Service to gauge their feelings on a cleanup of this site. Surprisingly enough, they were most agreeable, but only if I was able to get a release from the government for what they called 'government property on government land.'"

Forest Service officials knew about the wreck — it was one of several airplanes that dotted the mountaintops in the area — and were more than amenable to the cleanup. Many years ago, it was standard procedure for rangers or Forest Service employees to remove, dismantle, destroy or mark airplane wreckage that would otherwise tempt curious hikers into dangerous areas or fool pilots into thinking the area was the site of a recent aviation incident.

The "H" model Mustang escaped scrapping and destruction because of its remoteness, but either the military or the Forest Service had marked the fuselage and wings with a thick, latex-type paint to denote the aircraft was listed on crash rosters in case it was spotted from the air.

The Legal Office of the U.S. Air Force eventually released the airplane to the trio, but only after the crash paperwork was pulled out of the archives and reviewed, even though it had been forgotten for almost 40 years. Johnson said the request was examined by historians at Wright Patterson Air Force Base, who decided to release the rare Mustang because of "its extensive damage and the cost and effort that would be involved in the recovery."

"There were times when I thought we were going to be denied, and then everything started happening at once," Johnson said. "Within a couple of days, we had lined everything up, and we were hoping the weather was going to hold so we could operate the helicopter. It was touch and go. We were fighting temperatures in the high 90s in the mountains; not very conducive to operating helicopters."

A team of eight men worked several days at the crash site, all the while observed by the Forest Service to make sure the sanctity of the wilderness area was not disturbed. Johnson said 20 loads of parts were removed from the mountainside by a Bell JetRanger helicopter. "We used nets and nylon cargo slings and everything we had at our disposal to remove the wreckage," he said. "We used gloves to clean up small parts from rocks and trees, and the only thing we left behind was the Merlin. It is heavily damaged, but it's worth saving, and we hope to get it out [later], even if we have to disassemble it into components."

The wreckage was flown from the mountainside into a nearby valley, where it was loaded onto trucks and transported into a covered hangar in Scottsdale. The tension of removing some of the more unwieldy parts from the mountainside was broken when a family of chipmunks

Neil Jones in a Bell JetRanger lowers a wing of the recovered TF-51H Mustang. Robert A. Kropp Collection

(flying squirrels?) evacuated a wing in droves as the JetRanger pulled it airborne.

A post-recovery inspection uncovered perfect ammunition boxes in the gun bays, although all weapons had been removed during the training airplane's service days and the gun ports were faired over with stainless steel. Hydraulic lines were found to contain fluid, and radio sets still had tuners and crystals intact. The landing gear, which was still tucked up in one of the wings, could be fully extended; wiring in electrical junction boxes still smelled of fungicide preservative; data plaques, painted stenciling and pencil-marked inspection notes were still legible in the crumpled fuselage; and when a fuel cap was removed in a wing section, the intact rubber fuel cell still emitted a faint scent of aromatic fuel vapors.

Lt. Westlund's family members were presented with a North American Aviation-marked rudder pedal and some cockpit items when they visited the airplane. Ironically, the family had talked about the possibility of visiting Arizona to conduct their own search for the wreckage of the Mustang to put a chapter in their family history to rest.

Wing gun ammunition boxes look as good as the day Lt. Westlund took off from Fresno, Calif.

"We'll all never know just what happened during those last few moments as the Mustang fell to earth," Johnson said. "The accident report indicated Lt. Westlund tried to bail out of the stricken airplane, but got tangled up in the wreckage. His family told me that the plane allegedly had been back at the field twice before his fateful trip for repairs to the oxygen system.

"It was really a strange twist of fate. Lt. Westlund was a World War II combat pilot. His family showed me gun camera footage taken in some of his P-38 missions in Europe. The man survived some brutal aerial combat over Axis skies, and then he died in a freak mishap after stopping over in California to visit his sick grandmother and attempting to make an everyday cross-country navigation run. The accident report indicates he suffered oxygen deprivation and blacked out in flight. He came to with the airplane in a dive he couldn't bring under control, so he tried to get out of the airplane, and the Mustang just broke up around him. An odd twist of fate. We'll never know just what happened up there, and perhaps it's better that way."

Wing section and propeller after recovery.

Robert A. Kropp Collection

Although the trio had discussed keeping the Mustang for restoration, the rarity of the airplane and the great expenses borne in recovering the wreckage prompted them to part with it to a Southwest-based Mustang restorer. The rare Mustang was crated and shipped to Texas, where the new owner, who has asked to remain anonymous, plans to rebuild and mate the wings and tail section to a P-51H fuselage currently under restoration. The fuselage of the wrecked fighter will yield many valuable parts to the restoration, and the composite aircraft will reportedly be registered with the Federal Aviation Administration using the serial number of Lt. Westlund's former mount. The owner believes the P-51H Mustang eventually will return to the skies.

Arizona's rugged Galuero Mountains also have claimed a BT-13 trainer, a Boeing B-50, and a B-24 Liberator over the years.

Robert A. Kropp

B-24 training flight ends tragically on Arizona mountainside

A burned and mangled BC-348 aircraft receiver, its variable capacitor still tuned to its final frequency, lies wedged between boulders on a remote Arizona mountain peak.

Rocks bear scars where four huge Hamilton-Standard propellers hit at open throttle and then stopped; the spider gears of one propeller hub are twisted and broken in testament to the brute force of impact.

Less than 200 feet away from the impact site, control cables clang against a section of aluminum aircraft skin. Flung from the airframe during the impact, this section still contains a fragile dome light with its glass lens and 24-volt light bulb still intact.

The boulder-strewn mountain peak is littered with 18 tons of airplane parts.

The crew of this B-24 Liberator probably never knew what happened on that fateful night of September 15, 1944, when their heavy bomber flew headlong into Kachina Peak of the San Francisco Peaks.

Designated a TB-24J, she was built by Ford at Willow Run and carried the Army Air Forces serial number 42-50890. A training ship assigned to the Army Air Forces Western Flying Training Center (AAFWFTC), the Liberator was stationed with the 3007th Army Air Forces Base Unit at Kirtland Field, N.M. (now Kirtland Air Force Base).

Unlike other war-wearies usually assigned training duties, this Liberator was somewhat of an oddity. As a trainer, this Liberator was a "new" ship — the airframe had accumulated a total time of 645:40 hours since it rolled off the assembly line at Willow Run.

Two of her Pratt & Whitney R-1830s were original. Engine No. 2 was the newest replacement, with 295:40 hours; No. 3 had 334:30. This Liberator sported a bare-metal finish and was painted only with the national insignia, aircraft numbers, and anti-glare black paint on the inside of the nacelles and in front of the pilot's windscreen.

Second Lt. Warren E. Crowther, Topeka, Kan., was assigned to fly the Liberator on a navigation training mission on that fateful Friday morning in September 1944. The New Mexico-based crew was going home. No

Debris field of TB-24J, 42-50890, atop Kachina Peak. Fuselage section and bomb bays are recognizable.
Robert A. Kropp Collection

doubt each man on board had made plans for the weekend — movies to catch in town, parties at the officers' or enlisted men's clubs, or perhaps even dates awaited the young men on Saturday night. On the darkened airfield, the flight engineer walked around the bomber with a flashlight for a visual check, while the other crew members climbed aboard. The pilot and copilot began their pre-flight check.

Ignition switches "on." Throttles cracked one-third open. Booster pump "on." Starters — "ACCEL" switch up; then "MESH."

Lt. Crowther started the #3 engine, then #4, then #2, and then #1. As his copilot monitored engine oil and hydraulic pressures, the pilot taxied the heavy bomber toward the run-up area. Batteries were switched to generators; radio sets were turned on. The cold bomber was coming alive. Turning the bomber downwind, Lt. Crowther called for a run-up check.

Superchargers set and locked. Wing flaps, 20 degrees. Auxiliary hydraulic pump and power unit off. Cowl flaps closed.

After the tower cleared the Liberator for takeoff, Lt. Crowther released the brakes and turned the bomber into the active runway — trim tabs set for takeoff. Propellers set at "High rpm." Fuel mixtures — auto rich. The pilot opened the throttles, and the mighty Liberator began its takeoff roll.

As the Liberator lifted off the tarmac and began its gradual climb into the dark morning sky, the copilot lifted the landing gear, which folded into place with an assuring "clunk." Lt. Crowther and his crew, student pilots Ray Shipley, Austin, Texas; Clyce H. McCelvey, Vernon, Texas; Charles W. McDonald, Oklahoma City, Okla.; and Patrick Pertuset, Portsmouth, Ohio; along with engineer James R. Hartzog, Silver Creek, Miss.; radio operator John A. Franke Jr., Inglewood, Calif.; and student engineer Hugh H. Brown, Gastonia, N.C., departed from Bakersfield, Calif., to Kirtland Field at 12:48 a.m.

This may have been a familiar type of mission for someone like Lt. Crowther. Having earned his pilot wings just nine months earlier, he had racked up 512 hours in Liberators, 90 hours alone in the last month. One-fifth of his total time in B-24s involved night flying, and he had spent more than 15 hours in the left seat of Liberators on nighttime training flights in August.

overcast at 6,000 feet. Occasional showers were reported by a weather officer at a neighboring Army airfield at Winslow, Ariz.

But the radio signal received at Ashfork was the last message from TB-24J, 42-50890. Approximately 15 minutes later, the airplane crashed against the west slope of the San Francisco Peaks, an elevation of 12,500 feet.

The heavy bomber missed clearing the peak by less than 500 feet. The positioning of the wreckage indicates the Liberator may have clipped a wingtip on a pine tree, cartwheeled into the boulders, and exploded on impact.

A fairly intact outer wingtip is the largest section of wreckage. The landing gear is still tucked up into a wheel well in a portion of the massive Davis wing.

Almost 50 years after the crash, military officials are still vague about the exact cause. The accident report is stored on microfilm records at the Air Force Inspection and Safety Center at Norton Air Force Base, Calif.

The report, completed September 23, 1944, states the cause of the accident simply as, "Collision in full flight with

Some of the pieces of 42-50890 that are strewn on Kachina Peak. Robert A. Kropp Collection

The horizontal stabilizer is recognizable in the debris. Robert A. Kropp Collection

While the pilot may have been very familiar with the B-24, he had limited instrument flying experience — just over four hours total time, with half of that accumulated in the previous 30 days.

Lt. Crowther gently banked the plane and set a course, following a familiar route out of California and across Arizona. When he reached his desired cruising altitude, the pilot reduced power and trimmed the aircraft for level flight. Crew members coupled their oxygen masks to airflow regulators and plugged their heated flightsuits into rheostats at their respective stations. To this point, their actions were routine. The crew settled back and prepared to accumulate their air hours and training time. At 3:02 a.m., the bomber reported in over Ashfork, Ariz., which was experiencing a light storm with clouds broken to an

object other than aircraft. Airplane collided with mountain while in full flight." Accident investigators determined the scene of the crash was 15 miles off course on the north side of the airway.

In Section 19 of the report, the condition of each crew member is marked "Fatal." Under the column "Use of Parachute," investigators wrote simply "None" eight times.

On September 19, 1944, Lt. Harry E. Campbell, wrote, "Investigation of the wreckage revealed no signs of engine malfunction. All four engines were delivering power at time of contact with the ground."

The Liberator was written off as a "complete wreck" and soon forgotten. With wartime aircrew training continuing at a hectic pace, mishaps such as this one occurred all too often.

Portions of the accident report are still listed as "Restricted" information, especially specific information placing blame on the accident or forensic information.

Why did the Liberator stray so far off course? Did a Pitot tube ice up and affect instruments, fooling the crew into believing they were higher than they actually were in a well-known airspace rife with tall mountains? Did a moment's carelessness claim the lives of eight men? What was the rest of the crew doing when the heavy bomber struck the mountainside? Did the pilot see the mountain looming toward him and, at the last minute, heave on the controls and try to pull up? What did each man think in his final milliseconds of life?

No one, not even the military, can answer all of these questions. But when you're standing at 12,000 feet on a craggy mountain with your lungs straining from the thin air and your legs aching from the arduous uphill climb, the mind wanders. The mental picture of a solid, heavy bomber traveling invincibly at 175 mph is quickly dispelled by glancing out over the twisted graveyard of aluminum in front of you. In a battle between airplane and mountain, the mountain always wins.

Vincent P. Murone, chief of the Reports Branch at Norton, said portions of the report are not releasable and have been redacted. "These portions contain the opinions and conclusions of the safety investigator as to the cause or recommendations for mishap prevention purposes," he said. "Investigators are promised that their opinions and recommendations will be kept confidential as privileged safety information to encourage free and frank expressions of ideas. No time limit is placed on the promise of confidentiality; therefore, to ensure the continued confidence of current safety investigators in the safety privilege, we do not release opinions and recommendations even from these old safety reports."

Each year, thousands of people spend countless recreation hours in the shadow of the San Francisco Peaks, located about 10 miles north of Flagstaff. The mountains are a popular ski attraction during the winter and offer cool relief from the blistering triple-digit summer temperatures of the Phoenix metropolitan area. Despite the popularity of this mountain chain, the wreckage of the Liberator lies generally undisturbed because of its inaccessibility.

The "old timers" in Flagstaff who do know about the Liberator wreck at the top of the mountain often refer to it as "the B-17." They are partially correct — there is a B-17 wreck on the mountain top, but it is on another side. The confusion is easily understood considering the unusual circumstances. Less than one week after the Liberator crashed, a B-17G, which was on a training mission from Kingman Army Air Base, suddenly caught fire in flight and plummeted into Bear Jaw Canyon. All crew members on board the B-17 were killed.

Remains of prop hub assembly of 42-50890.
Robert A. Kropp Collection

A memorial plaque for the Liberator and her crew has been placed at a ski lodge at the base of Humphrey's Peak Trail. The memorial was erected in November 1988 by Mesa, Ariz., resident Richard Johnson and a local Boy Scout troop. The plaque reads, "In the early morning hours of Sept. 15, 1944, a U.S. Army Air Force B-24 bomber on a night training mission crashed into the San Francisco Peaks, killing all eight crewmen. This memorial is dedicated to those young American airmen.... For the cause of freedom, they gave up their lives." Before proceeding to the crash site, spend a moment of silence in memory of the crew.

The best way to get to the wreckage is to take the Humphrey's Peak Trail, nothing more than a little path through the densely wooded forest. A little less than two hours into the hike, the trail abruptly makes a sharp turn at a small clearing in the trees. The steep trail continues to the crest of the peak, an additional two miles. Off to the left is a massive, boulder-strewn rock slide. To get to the wreckage, leave the trail and carefully proceed across the rock slide and continue up to the top of the mountain.

The last mile up to the crash site is the most difficult because of the steep grade and the boulders, many of which are loose. Extreme caution should be exercised when on the rock slide, as a misplaced step could result in severe injury.

The remains of a Pratt radial engine will be found about a half-hour after crossing the rock slide. The powerplant, which broke free of the wreckage, separated mid-crankcase, and cartwheeled down the mountain, can be found propped up against a large fallen tree. The crankcase still contains trace amounts of oil, despite having been stopped while in full flight by propellers hitting rocks and trees and the passage of nearly a half-century.

The actual crash site is less than a half-hour from this

engine. Hikers will know they are getting close when they start seeing small pieces of aluminum.

The main wreckage is strewn over several hundred feet, much of it imbedded in the rocks. The force of the impact triggered a rock slide; the explosive sounds of the impact and the rumbling boulders must have reverberated down the quiet canyon on the morning of the incident.

A trip to the remote crash site is best done during June and July. Snow can still be found on the ground at the crash site in June. Despite triple-digit temperatures in the Phoenix metropolitan area less than three hours away, temperatures on the rocky peak can drop to the mid-40s at night. Appropriate dress is strongly recommended. The hike should be avoided in mid-to-late August because of Arizona's monsoons. Last year, two hikers were struck by lightning while climbing Humphrey's Peak Trail, one fatally.

The author has personally seen sudden storms roll in unexpectedly while visiting the crash site in August, and the aluminum-littered landscape draws lightning like a magnet is drawn to steel. A pair of thick leather gloves, sturdy hiking boots, and extra drinking water should be considered mandatory items for making this trek; the gloves are necessary not only for sorting through the razor-sharp pieces of aircraft wreckage, but for protecting one's hands on the rock slide. Some of the terrain is traversed on all fours.

Reverence and consideration should be exercised when visiting any aircraft crash site. While this Liberator remains fairly untouched by the masses over the years, some souvenir hunters have damaged parts that could have been utilized in a Liberator restoration.

A section of wing has been needlessly split by someone wielding an axe, when access could have been gained by simply moving some large rocks and crawling into an opening near a fuel cell cavity. Access panels have been bent open with pieces of engine mount, when an extra few minutes with a Phillips screwdriver or a Dzus tool would have kept the panel itself intact.

One souvenir hunter, intent on taking a propeller blade, used a hacksaw to remove the blade at the shank. Another souvenir hunter must have spent hours disassembling a propeller hub and carefully placed the hardware removed from the unit in a tin can. The conscientious souvenir hunter went so far as to wrap the exposed gears with a piece of plastic and carefully placed rocks over the gears, still covered with 1944-vintage lubricants, to keep out debris and moisture.

Several propellers, albeit bent and damaged, have remained at the crash site for more than 35 years. Each year, another "mysteriously" disappears.

The Liberator's flight deck disintegrated upon impact. Small items — panel switches, pieces of instruments, twisted rudder pedals, wires, tubing and oxygen regulators — are all that remain of the cockpit.

Pieces of thick plexiglass and heavy chunks of casting are all that remain of the nose turret. Not a trace can be found of the Martin 250 SE upper turret; usually the armor plate and the ring gear can be discerned even after an explosive impact and fire. Some who have visited the crash site have speculated that the top turret was removed on this training bomber to free up more room to accommodate the student pilots in the typically-crowded Liberator flight deck. Others have suggested the Martin could have broken free from the airframe during the impact and may have cartwheeled for several miles down the rugged rock slide. The latter seems improbable, as many heavy sections of armor plate and exhaust turbochargers remain in the immediate impact area.

A damaged but fairly intact Consolidated tail turret remained at the crash site for more than 40 years. The turret — still containing the seat, gun mounts, gun feed motor drives, ring gear, and many internal parts — was last seen at the sight in the summer of 1991. It is hoped that those who remove items from these historic crash sites are doing so to aid other aircraft restorations or out of personal interest in warbirds and not for scrap purposes.

No armament or ammunition can be found among the wreckage. It is doubtful that the airplane carried any weapons during its final navigation training assignment.

While the military removed the remains of the perished crew members from the crash site just days after the 1944 incident, some personal artifacts still remain in the wreckage. A shearling-lined, high-altitude flying boot was discovered a few years ago, in addition to several pairs of shoes, headsets, shirts, pants, pieces of flying jackets, and burned parachutes.

A damaged pair of sterling silver pilot wings were uncovered in the humus where the cockpit section impacted and burned, as was a dog tag belonging to Pvt. Brown. The dog tag was returned to the family of the perished crew member during a brief but emotion-filled ceremony of the memorial plaque dedication several years ago, and the other personal artifacts were returned to the earth not far from the site.

One of the outer wing panels is still clearly marked "250890." If you want to move it for a photograph, it takes at least two people to turn the wing over. Don't try to do it alone — it's too heavy and the rock base is unstable. When you're through photographing the wing and the numbers, turn it back over again to protect it from the sun, rain and snow.

The crash site is located in the Coconino National Forest. The forest service discourages the removal of items from the area.

Robert A. Kropp

Endeavour: Aussie airliner crash site revisited

Capt. Bruce N. Dickson departed Honolulu in command of a British Commonwealth Pacific Airlines' DC-6 on October 28, 1953. Known as Flight 304/44, Dickson was bound for San Francisco with eight crew members and 11 passengers. Sadly, Dickson, his highly experienced crew, and his passengers became involved in the worst air disaster to occur in San Mateo County.

British Commonwealth Pacific Airlines was founded on June 24, 1946, to operate trans-Pacific passenger and mail carrying flights from Australia and New Zealand to the United States and Canada. Share capital to start the airline was raised from the Australian government (50 percent), the New Zealand government (30 percent), and the United Kingdom (20 percent). Commercial service began in September 1946 with a bi-weekly flight from Sydney, Australia, to Vancouver, B.C. Auckland, New Zealand, was added as a departure point, and San Francisco became the U.S. destination. The Sydney/San Francisco flight was known as the "Southern Cross" route.

The first flights were made on Australian National Airways DC-4s under a charter agreement with BCPA. By 1949, BCPA had acquired three DC-4s and was bringing the new pressurized DC-6 on line. BCPA purchased four DC-6s. Each of the sleeper-equipped Royal Mail airliners was named after one of the ships (*Resolution, Endeavour, Adventure, Discovery*) commanded by Capt. James Cook (1728 - 1779) on his voyages of discovery in the Pacific.

Flight 304/44 was being flown by BCPA DC-6 *Resolution*, registered VH-BPE, the jumbo jet of the 1950s. *Resolution* was a Douglas-Long Beach-built DC-6, serial number 43125, completed in 1948. On October 28, 1953, she had 5,904 airframe hours since her last major overhaul. The DC-6 was a large aircraft, measuring 100 feet 7 inches in length, with a wingspan of 117 feet 6 inches. This DC-6 was powered by four Pratt & Whitney R-2800-CA15 engines driving Hamilton Standard propellers.

The flight originated at Sydney, where the following passengers had boarded: Reginald F.K. Eastoe, 50, an engineer from Norfolk, England; William J. Cox, 52, a shopfitter from Neutral Bay, Australia; John W. "Jack" Butterworth, 52, and Cyril G.H. McDonald, 60, engineers from the Victoria State Railway sent to the United States to study U.S. railway technology and to purchase a locomotive; Capt. Paul O. Olsen, 39, a master mariner from Cardiff, Australia, sent to the United States to take command of a new merchant ship; Bernard R. Tischler, 31, a chief passenger clerk from Brighton, Australia; John K. Briscoe, 45, an insurance executive from London, England; William Kapell, 31, a well-respected concert

Douglas DC-6 VH-BPH was a sister ship to VH-BPE Resolution, which crashed on Kings Mountain on October 29, 1953.
Photo courtesy John Wegg

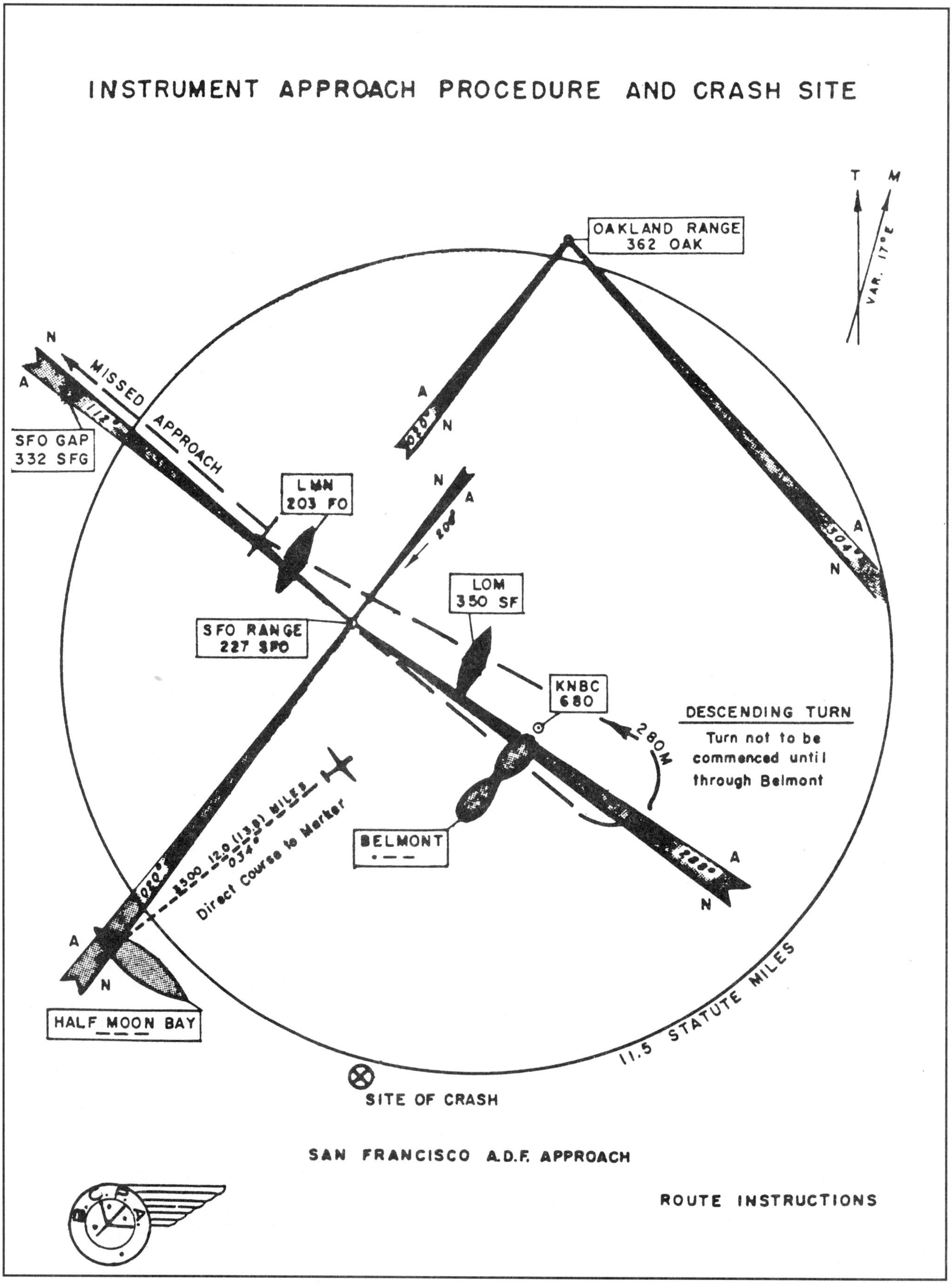

Diagram from the Civil Aeronautics Board showing the crash site of VH-BPE in relation to the approaches to the San Francisco Airport and the Half Moon Bay Fan Marker. Ed Davies Collection

pianist from New York City returning from a concert tour of Australia; and Janos Feher, 39, a watchmaker and citizen of Hungary, and his 7-year-old son, John, who had been stricken with polio, traveling to New York to see a medical specialist.

Resolution departed from Sydney with stops at Nadi, Fiji Islands, and then on to Canton Island and Honolulu, where a routine crew change took place. Capt. Bruce N. Dickson, 34, assumed command of Flight 304/44. He was a 10,696-hour pilot, with 3,624 hours on instruments and 4,136 hours in type. His crew on October 28, 1953, consisted of first officer Frank A. Campbell, 28, who held a third-class air transport rating with 3,896 of his 4,782 flying hours in the DC-6. George R. Murtagh, 28, navigator, had 2,482 hours in type, almost in exclusively trans-Pacific flights. Walter B. Knight, 34, purser, had 1,008 hours flying time. Vernon A. Walker, 32, radio officer, had 5,645 hours, 2,645 of which were in type. All of the crew members were Australian, except for Charles N. Cattanach, 31, flight engineer. Cattanach, a Canadian, was one of the first flight engineers hired by BCPA and had 3,118 hours, all in the DC-6. Rounding out the flight crew were hostesses June F. Elder, 27, and Amy K. Lewis, 32, both unmarried.

Mrs. Jean Chiverton, 65, of Vancouver, B.C., boarded the flight in Honolulu, becoming the eleventh passenger.

The estimated flying time from Honolulu to San Francisco was nine hours, 25 minutes, with sufficient fuel on board for 12 hours, 53 minutes. The gross takeoff weight was 90,166 pounds, 5,034 pounds below maximum gross takeoff weight.

Flight 304/44 departed Honolulu at 10:59 p.m. local time on an IFR flight plan to San Francisco. Once airborne, they requested to leave the tower frequency. Three minutes later, Dickson called the tower and said, "We're having a little trouble with one of the props." Permission was granted for a return to Honolulu, and the IFR clearance was canceled. At 11:05, Dickson reported, "The relay is working okay now; desire to continue flight." The IFR clearance was reinstated, and the flight proceeded towards San Francisco.

The normal radio communications were sent from the plane while en route, and Dickson estimated his arrival at San Francisco to be 8:40 a.m. PST.

At 8:21 a.m. PST, Western Air Defense Force Radar identified VH-BPE. VH-BPE was cleared "to the San Francisco ILS Outer Marker via the Half Moon Bay Fan Marker direct to the San Francisco Outer Marker. Maintain at least 500 feet above all clouds. Contact San Francisco Approach Control after passing Half Moon Bay Fan Marker. Cloud tops reported in the Bay Area: 1,700 feet." After crossing the coast at the Half Moon Bay Fan Marker, the standard approach directed the aircraft over the San Francisco Airport followed by a right turn for an approach to Runway 28. Capt. Dickson had flown this route numerous times and was well aware of the procedures.

Eighteen minutes later, *Resolution* advised that they were over Half Moon Bay, 500 on top, and listening on 278 kc. Approach Control responded with the following clearance: "Cleared for an ILS approach to the airport, Runway 28, wind west 15; cross the Outer Marker initially at least 500 on top, report when inbound; ceiling 1,200, visibility nine, altimeter 30.14." Unique to the microclimates of the San Francisco Peninsula, the coastal range was covered by a tree-clinging fog.

At 8:41 came the last transmission from VH-BPE, "Air Pacific Easy (VH-BPE's radio call sign), Roger, southeast, turning inbound." A controller quickly added, "Check passing the ILS Outer Marker inbound." *Resolution* was never heard from again, as all subsequent radio calls went unanswered.

Ray Brooks, now living in New Zealand, was the BCPA station manager at San Francisco on October 28, 1953. He had arrived for work at 7:30 a.m., in plenty of time to meet *Resolution.* Brooks worked closely with United Airlines, which was BCPA's handling agent at San Francisco. At 8:45 a.m., Brooks was informed that radio contact had been lost with *Resolution.*

Soon after the loss of radio contact, a Coast Guard PB-1 (B-17), piloted by Lt. Cmdr. D.M. Reed, was dispatched to search for the wreckage. Reed and his crew spotted the crash site at 10:10 a.m., one hour and 24 minutes after the crash. *Resolution* had crashed on the wooded southwest side of Kings Mountain, west of Redwood City, two miles from the Skeggs Point turnout on Skyline Boulevard.

The first on the scene was Belmont Police Officer James Winters, who discovered a completely demolished aircraft that had caused three forest fires. He noted correctly that none of the 11 passengers or eight crew could have survived. He was followed by the first rescue team led by Deputy Sheriff Joe Kimball, who was accompanied by *San Mateo Times* reporter Virgil Wilson and photographer Jack Russell. The rescue operation was coordinated by Redwood City Police Chief Stanley Wood. Paul Jensen, the newly elected coroner, had been in office less than 30 days, and this was to be his first major case. James Alteri was one of the first Red Cross volunteers at the scene.

Upon assessing the situation, bulldozers were called in to cut a road and clear a base of operations for the rescue and recovery operation. Once this base was established, San Mateo County Red Cross volunteer chairman, Liz Mack, organized canteen operations to feed the firefighters and Red Cross workers.

Mack, now living in San Mateo, said, "We used about 25 to 30 volunteers, nurses aides, a motor service, and a canteen service. We had a phone line set up within hours."

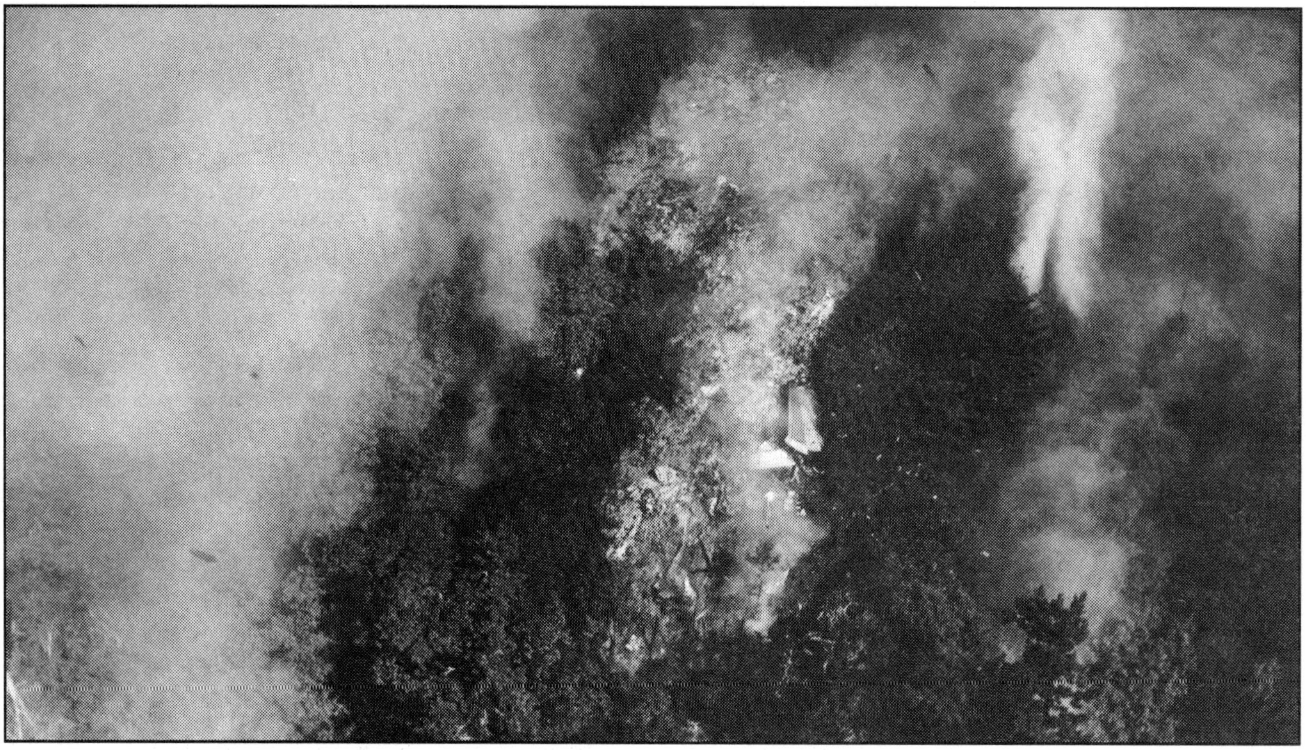

Aerial view of the crash site on Kings Mountain. Photo by Reginald McGovern

She also obtained potable water for the first rescuers from Vic and Alberta Brock, owners of the Kings Mountain Inn, a local restaurant at the time.

As soon as he received confirmation of the crash site, Brooks asked for the United crash team to be sent from Denver. He then left the airport to set up and coordinate the recovery efforts from the Redwood City National Guard Armory, which was serving as a temporary morgue. Brooks ensured that the mail was recovered, attempted to identify the bodies of the crew, and dealt with the insurance company representative who was on site.

The impact and resulting fires rendered most of the bodies unrecognizable. Coroner Jensen called in two physicians, Dr. A.C. Ladd and Dr. Seth Smott, and a dentist, Dr. Donald Allen, to assist in the identification of the bodies. These three men helped form the backbone of San Mateo County's first forensic investigation team.

The Civil Aeronautics Board investigated the crash and found that the aircraft was almost entirely destroyed by impact and the ensuing fire. The left wing first struck a tree at an elevation of 2,020 feet MSL, severing 13 feet 4 inches of the wing inboard from the tip. The main wreckage area was located at an elevation of about 1,950 feet.

Examination at this site revealed that the landing gear was down and locked, and the flaps were extended to between 15 and 20 degrees. All propeller blades were found, and from the blade positions, domes and hubs, it was determined that they were set between 29 and 37 degrees at impact. The propeller governors from engines two and three were found, and they indicated settings of about 2,400 rpm.

The CAB report states, "As far as could be determined from an examination of the damaged engines and components, there was no indication that a malfunction or failure had occurred prior to impact." The probable cause of the accident was given as, "The failure of the crew to follow prescribed procedures for an instrument approach." After seeing the location of the wreckage versus the Half Moon Bay Outer Marker, it is assumed that the aircraft crossed the coast south of the marker and was making a direct approach to Runway 28.

When BCPA was liquidated, the route was taken over by the Australian airline Qantas.

This 13-foot section of wing is the largest piece of VH-BPE at the crash site today. The point of impact is directly north on the opposite canyon wall. Photo by Ed Davies

VH-BPE *Resolution* Today

As far as known aircraft wrecks go, this has got to be the most accessible. The wreck of the *Resolution* is on land belonging to the Mid-Peninsula Regional Open Space District in the El Corte de Madera Open Space Preserve. Since the area is a park, all artifacts recovered remain the property of the Mid-Peninsula Regional Open Space District. Items recovered and now in the possession of the district include a three-ring notebook, a coroner's glove, suitcase hinges, an aircraft specialty tool, switches, clamps, and cockpit and side window glass. The list is long.

El Corte de Madera Open Space Preserve is open to the public from dawn to dusk. No motor vehicles or pets are allowed. Mid-Peninsula may be contacted at (415) 949-5500 for more information.

There are two ways to get to the wreck site. From Highway 92, drive south on Skyline Boulevard until you come to the Kings Mountain Country Store. Load up on sandwiches, chips and drinks and talk to Ken Fisher, the recognized historian of the Kings Mountain area. Check out his mini-museum and ask him for the free map to the El Corte de Madera Open Space Preserve. Someone made the map years ago, and Mid-Peninsula distributes them free. At one time, someone marked "airplane wing" on the map, but no one knew its significance. After collecting provisions and information, head south to Skeggs Point.

The second route is to drive west on Woodside Road from either Highway 101 or Highway 280 to Skyline Boulevard. Drive the two or three miles north to the parking area at Skeggs Point, walk across the road (to the west), and continue north about 100 yards to the gate at the beginning of the trail to Vista Point. Vista Point is about one and a half miles from Skyline Boulevard, and the trail is marked with signs and mileage estimates. A light-weight, long-sleeved shirt, good shoes, sunglasses or some form of eye protection, and a couple quarts of water are recommended for the off-trail walking to come later.

Vista Point is the area cleared by the bulldozers in October 1953 to serve as the base of operations for the recovery effort. From Vista Point take the trail west. About 10 feet from the end of the trail, head through the brush to your right (north). You know you are in the right spot if you see a white mannequin arm hanging from a tree. How it got there no one knows.

About 300 feet to the north on the down slope, you should see the 13-foot section of wing, which has the aircraft's registration, VH-BPE, painted in blue on the underside. After you have looked at it, please turn the wing back over, registration side down, to preserve the paint.

Getting to the wing is easy. However, if you are not adventuresome or willing to get dirty and cut by manzanita bushes, head back with the satisfaction that you have found the wreck of VH-BPE.

At the bottom of the ravine, you will begin to find large pieces of the airplane. If you are brave, venture up the opposite side of the ravine. Where the trees end and the manzanita bushes begin was the point of impact. A large amount of identifiable debris still exists — a radio rack, an ash tray, you name it. The ground is just littered with airplane parts.

Climbing to the top of the ridge where the manzanita is very thick is not recommended, and for those who might think, "there's got to be a road just over the hill," there isn't. There are a number of small pieces of fuselage, and the firewall section of an engine can be seen, but the pain of getting to them is not worth the reward. Return to the trail by hiking back to the wing and coming out the way you entered. It's the short, easy way.

Ed Davies, Nick Veronico

The BCPA Fleet

Name/ Registration	Douglas Constructor's Number	Delivered To BCPA	Disposition
Resolution VH-BPE	43125	Nov. 26, 1948	Crashed Oct. 29, 1953
Endeavour VH-BPF	43126	Nov. 19, 1948	Crashed Java, Jan. 1, 1971
Adventure VH-BPG	43127	Dec. 17, 1948	Air Tanker - N20MA, a longtime resident at Sonoma County Airport, with T&G Aviation, Chandler, Ariz.
Discovery VH-BPH	43128	Sept. 30, 1948	Air Tanker - N90MA, a longtime resident at Sonoma County Airport, with T&G Aviation, Chandler, Ariz.

Sole Surviving Martin B-10 Rests In California

Nose turret from the San Pablo Bay B-10. It is thought that this turret went to the U.S. Air Force Museum, Wright-Patterson Air Force Base, Ohio, in the early 1980s.
Photo by Gary Larkins

Martin B-10 in San Pablo Bay, at the end of the Hamilton Army Air Field runways. Photo by Jerry Knight

Bell P-39 Airacobra recovered from Canadian lake

Factory-fresh Bell P-39 Airacobra ready to go to war. Warren M. Bodie Collection via A. Kevin Grantham

During World War II, the United States lend-leased the Russians 4,924 Bell P-39 Airacobras. However, 178 of the P-39s that left the factory to serve with the Soviet air force did not reach Alaska. An aircraft wreckchaser's dream come true, somewhere between Buffalo, N.Y., and Nome, Alaska, lie the remains of these fighters.

The 25th Ferrying Squadron, 7th Ferrying Group of the Air Transport Command (ATC), was responsible for delivering lend-lease aircraft to Nome, where the Russian pilots would then fly the aircraft to the battlefields in Europe.

Two pilots from the 25th Ferrying Squadron took off from Fort Nelson, B.C., on the morning of December 6, 1943, in two Russia-bound Airacobras. Second Lt. Hal Coons and his wingman, 2nd Lt. Delos R. Carpenter, were airborne at 11:05 a.m. MWT. Their next stop was Whitehorse, Yukon Territory, with a final destination of Ladd Field in Nome.

Lt. Carpenter was flying P-39Q-15, serial number 44-2485. He had logged 475 hours flight time while in the Army Air Forces, 40 of those hours in the P-39.

Once airborne, Coons and Carpenter climbed through the "high overcast" and headed toward Fort Nelson on course. One hour later, Coons decided to let down through a hole in the overcast. As the aircraft descended, they became separated.

After making a number of 360-degree turns in an attempt to locate Coons, Carpenter, who was now low on fuel, decided to belly in on a frozen lake. He jettisoned his drop tanks and landed the aircraft with only minor damage to the propeller, but the right wing had buckled. The aircraft broke through the ice, but remained afloat. Carpenter removed his survival gear, which consisted of a shovel, an axe, rope and smoke flares, and then headed to the lake's shore, where he managed to build a fire. At approximately 8:30 p.m., Carpenter made contact with Lake Army Airways.

The following day at about 3 p.m., Carpenter was rescued by a "skiplane." The Army Air Forces reported that the ice was so thin and the lake so short, fuel had to be drained from the skiplane to get Lt. Carpenter out.

Lt. Coons had decided to set down because he had suffered a compass failure. Unable to locate Carpenter, he had followed a valley until he flew out from under the clouds, where he recognized his position as south of Ft. Laird. Coons returned safely to his point of origin, Ft. Nelson, to attempt the flight to Whitehorse another day.

Forty-seven years later, warbird restorer Gary Larkins of Auburn, Calif., set out in search of Lt. Carpenter's P-39. Larkins began by looking for the aircraft at the coordinates listed on the Air Force crash report. At the time of the crash, this area was "uncharted." The Air Force listed the crash location as 120 miles due north of the Smith River, so Larkins' first two trips were spent searching all of the lakes north of the reported crash site. Although these first two trips were unsuccessful, he secured the necessary permits from Canadian MOT — for the recovery of this and two other P-39s — in anticipation of a successful recovery of this aircraft.

On Larkins' third trip, he located 44-2485 in Carpenter Lake, 240 miles from Watson Lake, Yukon Territory, Canada. The aircraft was recovered on July 19 and 20, 1990, at the tail end of a 10-day trip into the Canadian wilderness.

Upon arriving at this particular lake, Larkins and his crew immediately knew that they had found the crash site. The area was littered with 55-gallon drums, the remnants of tents, wood stoves stenciled with U.S. Army Corps of Engineers, and a medical kit. George Carter and Wayne Lloyd, both former U.S. Navy S.E.A.L.s, worked with Larkins on the recovery effort. They are also the same divers who dove through two inches of ice to locate a PBM Mariner in Chesapeake Bay, only to find out that the aircraft was not worth recovering.

Larkins, Carter and Lloyd arrived at the lake in a DeHavilland Beaver fitted with floats. Aside from the regular camping gear, they took a rowboat, six dive tanks, a pair of lift bags, tools and a sonar scanner. Because of space limitations, they could not get an outboard engine on board the Beaver, but a set of oars fit nicely.

Their first two days at the lake were spent rowing the scanner the length of the lake. Once contact was made with the aircraft, they began diving on the Airacobra. They placed a buoy on the fuselage and then began to search for other parts underwater.

The nose of the aircraft was at a depth of 60 feet, and the tail was at 30 feet. While the aircraft was underwater, the divers removed the empennage. Carter and Lloyd had to work fast as they only had six air tanks, two of which were expended in the removal and recovery of the tail. Larkins raised the tail to the surface with a rope and then towed it to shore stroke-by-stroke with the rowboat. On shore, the horizontal was removed and readied for transport. Next they raised the airframe with lift bags. Again, using only the rowboat, Larkins towed it into shallow water, where they rigged it with a harness so that the helicopter could pick it up. Once rigged, the air was let out of the bags, setting the P-39 back on the lake bottom.

Former U.S. Navy S.E.A.L. divers, George Carter, left, and Wayne Lloyd, near the Airacobra's tail section after recovery.
Photo by Gary Larkins

After finding the P-39, Larkins radioed for the helicopter. It took three days for the Bell 205 to reach the wreck site at the right time of day. Once over the lake, a diver in the water coordinated with the helicopter pilot to pick the P-39 up. It took an hour to get the hook to the diver and onto the airplane. After this hour of maneuvering, fuel for the helicopter was getting critical. Once the hook was secured to the P-39, the fuselage and wings were lifted onto shore, where buckets full of mud were removed and the harness was double checked while the helicopter pilot took a well- deserved rest.

When the Army Air Forces salvaged the aircraft, they removed the engine, cockpit instrumentation, and the landing gear, flying those items out to Ft. Nelson. In the process of salvaging the P-39, the Army Air Force removed numerous pieces, setting them on the ice near the airplane. When spring came, the ice thawed, and all of the parts left behind sank straight down to a depth of 60 feet. Divers Carter and Lloyd recovered a number of inspection panels and the cowling, located the bent prop, and brought a number of other small parts to the surface.

Netting was rigged over the fuselage and wings to foil the lift, and a drag chute was deployed for directional stability during the airlift to Watson Lake. Once the P-39 was ready to be transported, the Beaver was dispatched to drop fuel at two locations while en route to pick up the divers, their gear, and the P-39's empennage. Transporting the P-39 over the 240 miles to Watson Lake took nine and one-half hours. The Bell 205 was only able to make 60 knots and at times had to clear mountain ridges as high as 14,000 feet.

Once at Watson Lake, the wing was demated from the fuselage, and all of the aircraft's components were packed into shipping containers. The P-39 is now awaiting export to the United States.

P-39Q, 44-2485, is in excellent shape. Both the Bell data plate and the Russian export data plate are still in the aircraft, and the Russian red star is still visible on the empennage and the wings. An Allison engine has already been acquired, and Larkins plans to recover two more P-39s to provide enough parts for a flyable restoration. Larkins' facility in Auburn, Calif., will conduct the restoration of 44-2485. Once the restoration has been completed, the P-39 will be flown and displayed in British Columbia.

Nick Veronico

Lifting the fuselage, center section, and wings back into the air after more than 47 years.

Photos by Gary Larkins

Bell P-39Q Airacobra, serial number 44-2485, on a trailer at Watson Lake. The helicopter flight from the crash site to Watson Lake took more than nine hours and covered 240 miles in the air.

In January 1943, a B-24D Liberator, Army Air Force serial number 41-11887, crashed on remote Bassett Peak in Southern Arizona, but was not located until March. The Consolidated-built Liberator was stationed at Tucson, Ariz., and was commanded by Lt. John Wasson. All 11 crew members perished.

Robert A. Kropp Collection

APPENDIX I

Aircraft wrecks by state

This listing of known aircraft wrecks was taken from the Civil Air Patrol's list of aircraft wreckage as well as from the Department of the Air Force, Headquarters Western Aerospace Rescue and Recovery Center's Aircraft Wreckage Locator for California. I have taken the liberty of culling out the civil aircraft wrecks, except for those I thought were rare types or large multi-engine aircraft.

This is a partial listing and is in no way complete. It is a common practice to delete aircraft from the crash locator lists as they become overgrown and unrecognizable.

Aircraft Type	Date	Latitude	Longitude	Remarks
Alabama				
T-34A	01/09/84	3118N	08647W	Wreck probably removed
Navy TH-57	10/27/82	3148N	08638W	2 miles south of Greenville
USAF T-37B	06/26/72	3214N	08655W	Details unknown
Arizona				
F-4	06/24/70	3131N	10940W	Bu. 637609, unmarked
AD-7	——	3139N	11025W	Scattered, unmarked
T-33A-1-LO	06/26/69	3145N	11018W	USAF 52-9496, silver
F4U/FG-1	——	3157N	10917W	7,000 ft. level, N.Y. Reserve markings
B-24	——	3210N	11053W	Fuselage at 6,000 ft. level
F-4	——	3210N	11316W	Details unknown
B-29/B-50	——	3217N	11146W	Scattered, unmarked
F-86D	——	3227N	11040W	7,000 ft. level of Mt. Lemon
B-50	——	3228N	11016W	USAF 46-04, scattered, unmarked
TF-104G-LO	10/29/69	3228N	11238W	USAF 63-8466, silver, scattered
F-14	02/28/83	3229N	11357W	Wreckage probably removed
P-51D-10-NA	——	3233N	11017W	USAF 44-14697, Bassett Peak
F-104	05/28/66	3234N	11243W	Silver
F4U	01/24/66	3236N	11225W	Marked
F-8C	04/07/65	3240N	11406W	Unknown
F-100C-25-NA	04/24/64	3243N	11251W	USAF 54-2067, burned
F-4C	05/06/65	3243N	10851W	Details unknown
DC-3	——	3244N	11003W	N127G, silver with red trim
BT-13	——	3246N	11027W	Oak Grove Canyon, Air Force markings
Interstate	03/16/65	3248N	11208W	N37292
BT-13	——	3250N	11051W	Unmarked pieces, Putnam Wash
FM-2	——	3250N	10938W	Bu. 55096, Safford
F-86	——	3257N	11043W	Wreckage marked
F-86	——	3302N	11432W	Scattered, unmarked
T-37	02/27/70	3305N	11111W	Silver, marked with yellow

Aircraft Type	Date	Latitude	Longitude	Remarks
P-51	——	3306N	11259W	Scattered, unmarked, Citrus Valley
AT-6D-NT	——	3311N	11211W	Burned parts visible
P-51	——	3311N	11230W	In small gully
T-33	——	3317N	11057W	AF098 marked
F-80	——	3319N	11139W	Details unknown
F-86F-30-NA	——	3320N	11012W	USAF 52-5100, silver
F4U	——	3320N	11111W	Broken up, marked
F4U	——	3320N	11218W	N9409, marked
F-104	04/17/65	3320N	11218W	Details unknown
AT-6	——	3321N	11110W	Burned, marked
B-24	——	3324N	11432W	Burned, unmarked, Mule Wash
F4U	——	3330N	11116W	Broken up, scattered, Weavers Needle
F-100D-25-NA	02/26/64	3334N	11245W	USAF 55-3605, scattered
TBM-3E	——	3336N	11106W	N8037E, wreck probably removed
AT-6	——	3338N	11137W	Adams Mesa
B-18	02/17/70	3340N	10955W	N230C, Bonito Prairie, white
T-33	——	3342N	10945W	Air National Guard, unmarked
F-80	——	3343N	11310W	USAF 669, scattered
P-51	01/01/52	3345N	11340W	Tail intact
C-47	——	3349N	11101W	Tail visible, scattered
P-80C-5-LO	——	3351N	11229W	USAF 47-547, scattered, unmarked
B-17	——	3354N	10929W	Scattered
C-45	——	3355N	10934W	Silver with Air Force star
AT-6	——	3355N	11213W	413283, unmarked
BT-13	——	3402N	11127W	Tail intact
F8F	——	3404N	11152W	N5151, top of peak, burned
BT-13	——	3405N	11127W	White/red, scattered
B-25	——	3405N	11214W	Unmarked
F-100	——	3406N	11315W	Scattered
BT-13	——	3407N	11125W	Unmarked
AT-6	——	3412N	11222W	Details unknown
F-84	——	3413N	11222W	Details unknown
T-33	12/08/76	3414N	11126W	N29147, unmarked
F-104	03/03/66	3415N	11252W	Silver
F-84	——	3415N	11357W	N69911, scattered
P-51 Mustang	——	3420N	11100W	N4893, unmarked
T-33	——	3420N	11133W	Marked yellow
Timm	——	3425N	11235W	NC6298, unmarked
F-100	04/16/66	3426N	11127W	Scattered
F-100	04/10/70	3427N	11122W	Camouflaged markings, burned
AT-6	——	3430N	11155W	Marked
B-25	——	3430N	11244W	In draw near top of mountain
F-84F	——	3431N	11220W	Unmarked
T-33	——	3432N	11154W	Details unknown
F8F	——	3432N	11246W	Scattered, exploded on impact
B-25	02/04/77	3433N	11412W	Scattered, Crossman Peak
F-8	08/02/69	3446N	11209W	Green
UC-78	——	3456N	11307W	Southeast slope of Mt. Hope

Aircraft Type	Date	Latitude	Longitude	Remarks
P-51	——	3500N	11124W	Ashurst Lake area
BT-13	——	3520N	11208W	NC6737, unmarked
T-33A-5-LO	06/24/70	3540N	11225W	USAF 57-0766, silver
T-33 or F-89	——	3540N	11353W	Marked
P-38	——	3550N	11150W	NX5101, unmarked
AT-6	——	3551N	11229W	Partially covered
P-47	——	3552N	11223W	NC9475, scattered
F4U Corsair	08/18/83	3509N	11212W	Wreck probably removed
California				
P-51	——	3152N	11451W	Details unknown
FJ-4	08/24/61	3233N	11642W	Bu. 139290, burned, unmarked
C-46	——	3235N	11649W	Details unknown
FJ-4	03/27/61	3236N	11651W	V-39291, at 2,500 ft. level
SNB	——	3236N	11614W-1/2N	NC68228, marked
F9F-2	——	3238N	11650W	Bu. 125142, NAS San Diego on tail
F4U	——	3240N	11700W	Bu. reported as 81140
F6F-3	11/30/43	3240N	11641W-1/2N	Bu. 66141
B-52	——	3247N	11515W	Marked
PBY	——	3248N	11614W	Details unknown
DC-3	——	3248N	11624W	Unmarked
F4U-1D	——	3251N	11612W	Bu. 57325
A4D	01/09/61	3254N	11543W	Unmarked
P-39	——	3254N	11627W	USAAF 42-6695
F4U	——	3255N	11504W	Marked
F4F	——	3259N	11550W	Marked
P-51	——	3310N	11438W	Unmarked, listed as AF2120
F8U	06/31/63	3302N	11553W	Bu. 146977, marked
UC-78	——	3302N	11707W	NC44000, marked
F8U-1	02/17/60	3304N	11504W	Bulldozed over
L-19	10/20/63	3305N	11634W	USAF 40088
F8D	——	3310N	11720W	Details unknown
SNJ	——	3325N	11615W	Details unknown
F8U-2NE	09/03/64	3322N	11531W	Bu. 150881, scattered, unmarked
F8U	03/27/63	3322N	11648W	Scattered
AD	——	3322N	11731W	Marked
FG-1	——	3329N	11542W	Marked
F8U-1	08/03/60	3330N	11633W	Bu. 145504
FJ-3	——	3331N	11517W	Listed as NGG5814
F-80A-1-LO	——	3332N	11620W	USAAF 44-85208
TBF	1950	3332N	11747W	Underwater, not visible
T-33	09/18/64	3332N	11729W	USAF 52-9670
F8U	09/20/61	3333N	11626W	Burned, unmarked
AD-5	1960	3334N	11619W	Marked
SBD	——	3334N	11732W	Marked
F8U-1	11/15/62	3335N	11618W	Bu. 143728, marked yellow
B-24	——	3339N	11449W	Details unknown
AD	——	3340N	11745W-1/2W	USMC, unmarked

Aircraft Type	Date	Latitude	Longitude	Remarks
AD	——	3341N	11646W	Details unknown
PT-26	——	3342N	11637W	NC60506, unmarked
C-45	1948	3342N	11733W	Marked, yellow X
UH-1	04/13/67	3342N	11732W	USMC markings
BT-13	——	3344N	11635W	Details unknown
P2V	05/29/63	3344N	11734W	Scattered
BT-13	——	3346N	11637W	Details unknown
F9F	——	3346N	11726W	Details unknown
F9F	——	3346N	11740W	Details unknown
F3F	——	3347N	11743W	Bu. 5843, marked
P-38	——	3348N	11726W	Marked
C-47	12/14/61	3348N	11737W	Burned
PT-19	——	3349N	11658W	N48588, unmarked
F-86	——	3355N	11611W-1/2W	Details unknown
F-89	——	3358N	11743W	Unmarked
P-51	——	3400N	12000W	Listed as AF4542
AT-6	——	3401N	11545W	Details unknown
F9F	06/23/61	3402N	11601W	Listed as Bu. 147380
PT-19	——	3402N	11648W	NC56575
BT-15	——	3403N	11556W	USAAF 42-1833, marked
B-25	01/26/55	3404N	11648W	USAF 44-86805, marked
C-47	12/01/52	3406N	11649W	USAF 45-1124, marked
PT-17	——	3408N	11646W	Details unknown
BT-13	——	3408N	11711W	NC69500, marked
A-26	——	3410N	11735W	N717L, unmarked
BT-13	——	3410N	11717W-1/2N	Details unknown
B-26	——	3411N	11815W-1/2N	Details unknown
SNJ-5	1944	3412N	11652W-1/2N	Marked yellow
F-100C-25-NA	09/21/56	3412N	11711W	USAF 54-2027
F-100	05/27/64	3412N	11722W	Scattered, unmarked
C-47	——	3412N	11732W	Marked
SNJ	——	3412N	11759W	Marked
C-45	——	3412N	11816W-1/2W	Details unknown
P-38L-VN	——	3413N	11819W	USAAF 44-24918, marked
AT-6	——	3413N	11807W-1/2N	Marked
BT-13	——	3414N	11647W	Marked
A-26	——	3414N	11733W	Listed as 5488, marked
BT-13	——	3415N	11815W	Details unknown
B-24	——	3416N	11722W	Details unknown
P-51	——	3416N	11722W	Details unknown
F6F	——	3416N	11736W	Details unknown
UC-78	1952	3416N	11808W-1/2W	Unmarked, scattered
F-89	01/31/57	3416N	11822W	Scattered
C-45	1945	3417N	11725W	Details unknown
T-33	02/10/59	3417N	11740W	USAF 56-3683, scattered
F9F-5	02/02/57	3417N	11742W	Bu. 125559
F6F (2)	——	3417N	11739W	Marked
C-46	——	3418N	11739W	Reported as 15190, unmarked

Aircraft Type	Date	Latitude	Longitude	Remarks
B-24	——	3419N	11809W	Reported as 42-90434, marked
H-19	1954	3422N	11744W	Marked
C-47	09/21/61	3420N	11836W	N16002, marked
F-86	10/24/56	3421N	11722W	Details unknown
P-80A-1-LO	——	3421N	11729W	Reported as 51371
AT-6	05/14/60	3421N	11825W	N2801G, unmarked
F-84F	02/27/57	3422N	11428W	Reported as AF065, bailout
B-57E-MA	05/03/63	3422N	11924W	USAF 55-4296, scattered
T-34	12/31/64	3423N	11755W	N5506V, yellow with black trim
B-25	——	3423N	11814W	Scattered, silver
C-46A-10-CU	——	3423N	11752W	USAF 41-12363, marked
SBD	——	3424N	11822W	Marked
T-33	——	3425N	11910W	Details unknown
AT-6	——	3427N	11859W	Details unknown
F-86 (2)	09/04/56	3428N	11803W	61338 and ANG 48313, unmarked
P-51	——	3430N	11900W	Details unknown
C-47	——	3432N	11841W	Details unknown
H-34	07/08/64	3432N	11859-1/2W	Bu. 145714, red & black #5 on side
BT-13	——	3432N	11859W	Burned, unmarked
DC-3	——	3432N	12001W	NC63107, marked
Curtiss Jenny	——	3432N	12001W-1/2N	Details unknown
B-24	——	3434N	11943W	Scattered, unmarked
B-26 (2)	——	3434N	11735W-1/2N	AF677 & AF174, unmarked
F4F	——	3434N	12013W	Details unknown
Howard DGA	——	3435N	11905W	NC69965, marked
BT-13	1957	3435N	11905W	N63118
A-1	08/19/65	3436N	11621W	Reported as 35331, white
P-38	08/04/65	3437N	12012W	Burned
P-51	——	3438N	11719W-1/2W	AF978, may have been removed
Stinson	——	3440N	11923W	Unmarked, in river
DC-3	——	3441N	11858W	NC18645, unmarked
P-38	05/28/56	3442N	12009W	N504MH, removal attempted
A-4C	08/18/66	3443N	11617W	Details unknown
F4F	10/31/59	3443N	11907W	Reported as aircraft #259
T-33	12/23/65	3446N	11523W	USAF 53-5956, marked
AD-5	08/25/61	3448N	11755W	Reported as aircraft #33876
F-4	07/10/67	3449N	11722W-1/2W	65-0675, silver, burned, marked
C-47	——	3449N	11911W	Aircraft #KC345
Howard	11/18/61	3451N	11501W	N1441
P-51	——	3451N	11535W	N7369B, unmarked
F-4D-28-MC	05/18/67	3453N	11558W	USAF 65-0668, silver, burned
HH-43B	12/07/62	3455N	11830W	USAF 60-286, slated to be removed
F6F (2)	——	3455N	11900W	Mid-air collision
T-33	——	3455N	11916W	Navy, wing only, unmarked
AT-6	——	3455N	11956W	Unmarked
C-47	——	3456N	11532W	Reported as 112363, marked
B-26	01/31/56	3456N	11556W	Tail only, thought to be an A-26
UC-78	——	3456N	11837W	Reported as 298831

Aircraft Type	Date	Latitude	Longitude	Remarks
AT-10	——	3457N	11835W	Reported as 23512, unmarked
AT-6	——	3458N	12013W	Details unknown
BT-13	——	3502N	11831W	USAAF 41-1500, marked
BT-13	——	3502N	11833W	Reported as 63334
F4U	——	3502N	11850W	Reported as 1542
A-20	——	3502N	11829W-1/2N	Reported as USAAF 41-1564
P-51	04/04/51	3504N	11718W	Possibly removed
XB-70/F-104	06/08/66	3505N	11703W	Silver
F-10	02/11/64	3506N	11640W	Scattered
F4U	——	3512N	11823W-1/2N	Marked
T-34	11/18/60	3513N	12006W	N5507V, marked
F-104C	04/12/63	3515N	11610W	56-896, details unknown
C-45	12/16/61	3517N	11635W	N16020, burned
Aeronca	——	3520N	12038W	NC31994, unmarked
Howard	——	3522N	11825W	N67764, unmarked
UC-78	——	3522N	12037W-1/2W	UC69080, unmarked
HRS-2 (S-55)	——	3535N	11529W	Marked
P-38L-1-LO	——	3531N	11612W	USAAF 44-24790
PT-19	——	3528N	12059W	NC48885, marked
Single-Engine Fighter		3528N	11700W	3500 ft. level, silver wing, orange wingtip
Stinson	10/12/47	3533N	11531W	NC591000, unmarked
AT-6	10/26/45	3538N	11535W	AF2913, unmarked
KC-135	07/08/64	3537N	11647W-1/2N	Widely scattered, burned, visible
F-105	07/08/64	3541N	11644W	Partially removed
F-100	——	3542N	11657W	Marked yellow
PQ-14	——	3543N	11646W	USAAF 43-4439, unmarked
JRB-4 (C-45F)	02/03/49	3543N	11759W	Bu. 90550, unmarked
F-4C	02/20/67	3547N	11556W	Reported as 65-0656, marked
B-25	1944	3547N	11838W	USAAF 41-30140, marked
AD-5Q	04/11/61	3550N	12110W	#134988, burned
Stearman	08/02/57	3557N	12112W	NC5722B, unmarked
F-104	——	3557N	11831W	Scattered, turbine visible
AD-4	12/17/61	3606N	11641W	Reported as 42203
FG-1	06/13/47	3603N	11636W	Marked yellow
WACO	1947	3606N	11807W	NC17745, marked
Stinson	05/14/53	3607N	12100N	N9890, marked
UC-78	1946	3611N	11822W	NC1342, marked
PT-17	06/06/46	3612N	11820W	NC6454, marked
BT-13	1951	3613N	11806W	Wreckage is orange
F-100	08/18/58	3615N	11645W	Scattered
F-100	10/22/56	3618N	11643W	AF654, details unknown
A-4	10/15/65	3622N	11850W	White, aircraft #504
B-17	1942	3622N	12045W	Unmarked
F6F-5	1945	3623N	11813W	Bu. 71033
SA-16	——	3625N	11719W	Details unknown
A-20	——	3626N	12154W	Marked
C-45	04/08/64	3629N	11814W	Marked, 7 SW of Mt. Whitney
C-46	01/13/59	3634N	11843W	N1240N, marked

Aircraft Type	Date	Latitude	Longitude	Remarks
R4D (C-47)	02/02/45	3635N	11820W	Navy 39584, marked
SNJ	08/23/52	3642N	11821W	Navy 0994, marked
P-40C	10/24/41	3642N	11831W	USAAF 39-287, details unknown
U-3B	06/18/66	3643N	11823W	Blue/white, AF06081, visible
H-19	06/26/66	3644N	11821W	Bu. 130222, blue, submerged in lake
P-38	1942	3644N	12108W	Marked
B-24	10/20/44	3645N	11751W	Reported as 42-41444, marked
SA-16A	05/25/53	3654N	11829W	AF7173, marked
B-52B-BO	09/17/56	3655N	12000W	USAF 53-0393
F2H-3	08/02/56	3656N	11822W	Unmarked
F-86	11/23/54	3704N	12051W	Reported as 3766, unmarked
B-18	1940	3705N	11825W	"Army" on wing
TV-2	10/07/56	3706N	12151W	Reported as 8060
F9F-8	——	3707N	12138W	Bu. 146374
F3H-2	09/20/60	3708N	12012W	Bu. 143442, marked
F3H	10/07/61	3708N	12212W	Navy 3766, marked
C-47	——	3712N	12130W	White, marked
TV-2	04/14/55	3712N	12205W	Bu. 138055, marked
P-51	——	3714N	11851W	Details unknown
A3D	10/26/62	3714N	12055W	Possibly removed
F2H-4	02/22/59	3717N	12207W	Reported as 127614
A-4	07/08/65	3722N	12121W-1/2N	Navy NF461
F8J (2)	03/24/59	3727N	12134W	Bu. 144448 & 144451, wing section visible
PT-19	——	3729N	12141W	USAAF 41-25537, marked
PT-19	09/25/60	3730N	12220W	NC69086
T-33	01/11/65	3734N	12157W	Reported as 41506, scattered
F4H-2B	10/29/56	3734N	12144W	N125651, unmarked
A-4E	02/20/64	3736N	11852W	Bu. 150077
R5D	02/17/56	3737N	12156W	Bu. 39116
BT-13	08/08/48	3738N	12158W	NC57365, marked
T-28	02/17/55	3743N	12159W	Bu. 17602
A-4D	02/20/64	3744N	11904W	Navy 504, scattered
F-101	01/10/64	3747N	12205W	Reported as USAF 15438, burned
FJ-3	10/26/56	3751N	12212W	Bu. 135978
C-47	03/16/45	3751N	11912W	USAAF 43-3067, 9,000 ft. level, marked
UC-78	06/04/47	3745N	12038W	NC5458, marked
F9F-6	10/30/56	3755N	12210W	Reported as 12758
TV-2	06/29/56	3801N	12128W	Navy 750, unmarked
P-61	08/28/44	3804N	11926W	Known dynamited by USAAF late 1940s
C-45	——	3804N	12238W	Details unknown
P-39 or P-63	——	3804N	12255W	Details unknown
A-4	07/10/66	3807N	11935W	Navy 2789, scattered
H-34	03/23/65	3821N	11939W	Reported as 48935, intact
T-28	07/02/65	3823N	11204W-1/2W	N8321H
F4U	01/29/52	3825N	11933W	Bu. 82067, marked
F2H	08/25/55	3831N	12236W	Bu. 123344, unmarked
B-24D-20-CO	1942	3835N	12310W	USAAF 41-24262, unmarked
T-33A	02/19/56	3839N	12237W	USAF 51-9307, unmarked

Aircraft Type	Date	Latitude	Longitude	Remarks
Fighters (2)	08/26/54	3839N	12239W-1/2N	AF5254 & 5264, marked
T-33	04/12/61	3842N	12132W	Details unknown, burned
S-2	01/16/66	3855N	12014W	Bu. 136542, gray, scattered
Fighter	02/13/58	3855N	12332W	Bu. 135048, marked
Constellation	03/09/64	3900N	11953W	N86504, Paradise Airlines
C-46	Apr 1944	3901N	12056W	Marked
P-39	03/21/43	3901N	12222W	USAAF 42-9120, scattered
BT-13	10/25/50	3904N	12049W	N56639, marked
AT-11	12/17/49	3912N	12052W	AF9564, unmarked
BT-13	03/07/47	3913N	12033W	NC58661, marked
BT-13	01/13/47	3915N	12116W	NC62742, marked
Stinson	05/30/47	3916N	12101W	NC97999, marked
T-33	11/13/55	3925N	12020W	Reported as 4339, unmarked
T-33	02/27/57	3925N	12043W	Reported as 51-4495, unmarked
BT-13	12/03/45	3927N	12242W	Reported as AF9669, marked
TBM-3U	Aug 1960	3928N	12016W	NC109Z, probably removed
Navy Jet	——	3939N	12116W	Scattered, unmarked
BT-13	——	3939N	12251W	Scattered, burned
B-17F	05/30/43	3955N	12306W	42-5318, Leech Lake Mountain
B-29	1944	3957N	12144W	May be B-29A-50-BN, USAAF 44-61843
A-4C	08/19/66	4004N	12404W	Details unknown
L-20	01/12/59	4032N	12047W	Reported as 43-2086, marked
BT-13	11/13/47	4039N	12112W	NC58778, marked
B-26	1944	4059N	12250W	Details unknown, marked
B-26	——	4101N	12253W	Reported as 42-87186
BT-13	12/16/49	4107N	12126W	NC56865, marked
T-34	02/27/60	4107N	12144W	N6186C, unmarked
F6F	11/14/45	4110N	12230W	Covered, unmarked
TBM	06/06/47	4121N	12300W	Unmarked
F-101B-70-MC	05/09/62	4147N	12143W	USAF 57-0271
Colorado				
B-24	——	3715N	10510W	Marked, visible
T-29	02/25/62	3725N	10803W	91931
BT-13	——	3727N	10513W	Scattered
L-1049	12/15/65	3734N	10530W	N6914C, scattered, burned
L-5	01/15/44	3751N	10500W	AF8693
A-4C	09/14/69	3755N	10723W	Gray with black tail
AT-6	03/25/67	3830N	10603W	N6973C
C-47B-1-DK	09/24/56	3831N	10618W	43-48416
T-33A-5-LO	09/12/74	3840N	10539W	58-0528, silver, burned
L-19E-CE	08/16/51	3840N	10503W	56-2669
U8F	09/14/70	3843N	10505W	A23859, olive drab
B-25	04/29/58	3845N	10503W	29785
B-58	04/22/58	3849N	10503W	AF8102
C-47	10/14/47	3851N	10504W	AF3503, marked
B-17	——	3858N	10509W	Marked
C-49	12/01/45	3858N	10503W	Burned

Aircraft Type	Date	Latitude	Longitude	Remarks
C-46	12/07/63	4002N	10538W	N609Z, silver and blue
DC-3	01/21/48	4003N	10538W	N206, silver
F-100	01/30/67	4025N	10553W	USAF 54-2015
Unknown	——	4025N	10544W	Radial engine, stabilizer visible
Unknown	——	4025N	10555W	Military, human remains recovered
B-17	——	4029N	10526W	Marked
B-17	06/14/47	4031N	10538W	AF0891, marked
B-24	09/18/45	4035N	10707W	AF8549
C-46	11/16/63	4037N	10529W	N130N
A-7 Corsair II	12/10/81	4039N	10357W	Wreckage probably removed
DC-6	06/30/51	4045N	10544W	Silver
T-34 Mentor	08/01/62	4047N	10705W	N0828
Connecticut				
Cessna 150	——	4132N	07327W	Details unknown
Cessna 182	01/03/86	4140N	07235W	N3349R, white/gold, probably removed
Delaware				
A-7B Corsair	01/15/83	3341N	07856W	In water
A-7B Corsair	01/15/83	3341N	07856W	Wreck probably removed
Florida				
F-4	07/11/83	2439N	08023W	Crashed in water
Navy Corsairs	01/01/54	2527N	08051W	Details unknown
DC-3	———	2625N	08037W	Marked
DC-3	03/20/77	2634N	08111W	N101MX, white
F-4	09/07/71	2641N	08055W	Details unknown
DC-3	——	2658N	08034W	Silver wing and tail
DC-3	——	2705N	08135W	White
F-4	06/28/73	2735N	08117W	Camouflage
F-100C-20-NA	04/22/68	2738N	08109W	USAF 54-1914
T-33	——	2803N	08057W	Details unknown
Unknown	12/08/81	2804N	08006W	Crashed in water
A-5	10/04/69	2853N	08048W	Details unknown
F4U Corsair	——	2902N	08116W	Unmarked
F-101	12/14/71	2938N	08518W	Details unknown
T-33	——	3004N	08535W	Details unknown
F-101B-115-MC	12/04/75	3019N	08522W	USAF 59-0406, tail intact
T-33	12/22/82	3021N	08526W	Markings: 28N Tyndall AFB
T-28	03/11/68	3040N	08656W	Marked as 2W216
A-6 Intruder	02/12/82	3047N	08226W	USMC, Okefenokee Swamp
Georgia				
T-33	——	3135N	08337W	AF1964
P-40	——	3137N	08110W	Given as 0244
F-4	04/30/41	3426N	08442W	Probably removed
F-101	——	3438N	08433W	Details unknown
UH-1B	03/22/76	3447N	08405W	USAF 63-8762, green

Aircraft Type	Date	Latitude	Longitude	Remarks
Idaho				
RF-4C-26-MC	10/04/69	4223N	11603W	USAF 65-0889
C-140	11/26/53	4231N	11205W	C76661
AT-6	12/01/63	4313N	11528W	Visible, Bennett Mountain
T-33A-1-LO	09/06/62	4317N	11432W	USAF 51-17419, scattered
BT-13	02/31/46	4324N	11553W	C325
P-38	03/11/45	4325N	11535W	Unmarked
L-19	12/13/55	4336N	11526W	Intact, unmarked, Trinity Mountain
B-47	01/01/62	4413N	11605W	Burned
C-43	04/05/43	4430N	11510W	Cape Horn Lakes area, unmarked
C-47	01/01/58	4445N	11648W	USFS, aluminum, Advent Gulch area
C-45	07/09/65	4450N	11450W	Scattered
B-24	01/27/43	4509N	11550W	Loon Lake area, visible, unmarked
Taylorcraft	10/12/45	4511N	11451W	Visible, unmarked
L-5	06/28/63	4521N	11529W	58-995, yellow and blue
TBM	——	4537N	11353W	Marked
B-17	11/01/62	4702N	11640W	Details unknown
B-24	06/16/41	4737N	11700W	"4443," Rockford Bay area
BT-13	04/29/47	4741N	11629W	C56962
Iowa				
B-23	02/15/67	4212N	09045W	N2318Q, white with blue trim
Louisiana				
BT-13	——	3003N	08945W	Aircraft on its back
H-34	——	3005N	09027W	N46755
Maine				
F-101	11/14/67	4458N	07006W	Details unknown
TBM Avenger	05/18/72	4656N	06939W	Details unknown
Montana				
TBM	08/27/64	4537N	11353W	Unmarked
T-33	05/05/65	4701N	11101W	Tail number "079"
F-101	05/19/66	4802N	11149W	Silver
Nebraska				
A-26	09/18/65	4153N	10333W	Executive conversion
New Mexico				
B-25	——	3211N	10458W	Not visible from air
AT-11	——	3212N	10633W	Details unknown
B-25	——	3217N	10633W	Details unknown
Unknown	——	3225N	10558W	Military, silver
BT-13	——	3231N	10628W	Details unknown
C-140	——	3246N	10544W	On bombing range
T-34	08/04/81	3301N	10745W	Forest Service aircraft
T-33	05/26/81	3320N	10503W	USAF, 25 miles west of Roswell
OV-10	08/06/81	3328N	10708W	At the end of Elephant Butte
T-33	07/28/67	3336N	10412W	USAF, silver
F-4C-26-MC	12/06/69	3357N	10527W	USAF 64-0969, camouflage

Aircraft Type	Date	Latitude	Longitude	Remarks
B-17	——	3406N	10710W	Details unknown
A-70	——	3423N	10349W	Details unknown
T-37A-CE	02/23/69	3440N	10625W	USAF 57-2301, silver and black
AT-11	——	3446N	10626W	Details unknown
Unknown	——	3909N	10624W	N8414E, Navy
AT-6	——	3515N	10553W	Details unknown
C-47	——	3515N	10611W	Details unknown
B-24	——	3624N	10503W	Details unknown
Navy PBM	01/01/64	3629N	10528W	Marked yellow
C-47	——	3647N	10527W	Marked
B-17	——	3647N	10531W	Details unknown
B-24	——	3653N	10521W	Visible from air near peak
B-17	——	3655N	10830W	Details unknown
North Carolina				
F-11	05/20/69	3502N	07715W	Bu. 138639
T-34	11/19/73	3530N	07737W	Details unknown
F8F	01/01/64	3608N	07726W	USMC
Oklahoma				
F-105	09/13/80	3419N	09535W	Details unknown
Oregon				
A-6	09/20/73	4306N	12026W	Navy 28314, silver and white
A-6	05/31/73	4318N	12242W	Navy 52907, gray
F-101	04/06/65	4324N	12055W	Details unknown
DC-9	10/02/66	4515N	12158W	N9101, marked
Culver	12/08/68	4547N	12257W	NC4461
Pennsylvania				
T-34	10/23/82	4007N	07916W	Navy
Texas				
T-38	04/06/83	2951N	10041W	USAF 1591
S-2	05/23/65	3014N	10108W	V33111, gray
T-38A-50-NO	05/26/65	3027N	10129W	63-8244, scattered
B-24	——	3131N	10534W	Scattered
B-29	——	3146N	10446W	Scattered on west slope
Lockheed 12	——	3150N	10630W	Scattered
TV-2 (T-33)	——	3157N	10427W	Aircraft in gully
P-51	10/15/75	3159N	10053W	N651D
T-33	——	3234N	09725W	Details unknown
T-37	12/17/69	3256N	10112W	USAF 58-1913
F-100	02/09/69	3418N	10100W	Red, white, blue
Utah				
UC-78	——	3731N	11358W	Details unknown
T-33	——	3820N	10939W	Details unknown
T-28	07/02/65	3823N	11204W	Reported intact
F-4E-51-MC	12/03/77	3947N	11403W	USAF 72-0134

Aircraft Type	Date	Latitude	Longitude	Remarks
F-111	03/01/69	3955N	11245W	Details unknown
V-1 Rocket	——	4026N	11333W	Intact, unmarked
Mace Missile	——	4033N	11335W	Scattered
V-1 Rocket	——	4034N	11353W	Intact unmarked
T-34	11/22/65	4039N	11044W	N6826E, silver
Virginia				
A-6 Intruder	06/27/83	3636N	07725W	Wreck probably removed
FB-111	——	3656N	08153W	Wreck marked with white/yellow
S-2	02/12/69	3700N	08010W	Details unknown
T-28B	05/01/64	3756N	07854W	USMC Bu. 137731

APPENDIX II

U.S. Navy pilot history cards of aircraft accidents

The following Pilot History Cards of Aircraft Accidents are on 16mm microfilm reels. Each roll costs $15, checks payable to the Department of the Navy. They are available from:

Department of the Navy
Naval Historical Center
Naval Aviation History Branch
Building 157-1
Washington Navy Yard
Washington, D.C. 20374-5059

NAVY AVIATORS TELL OF EATING SHARK KILLED WITH PENKNIFE. (By the Associated Press)

Pearl Harbor, Hawaii, March 10. A gripping story of beating off sharks with bare fi[illegible] and of batting hunger and thirst while stark-naked and half-crazed came today from three Navy fliers who survived 34 days in stormy South Seas on a raft smaller than an ordinary bed.

From the time their land plane was forced down on the Pacific until Gene Aldrich, 24 year old radioman of Sikeston, Mo., mouthed through parched lips the startling words, "Chief, I see a field of corn," the three lived a story that would tax the imagination of [illegible]fiction writer. The cornfield Aldrich thought he saw was a coconut grove.

"Chief" is Harold Dixon, 42, of LaMesa, Calif., aviation chief machinists mate and pilot of the lost plane. The third of the fliers is Anthony J. Pastula, 24, of Youngstown, O., aviation ordnanceman. They traveled at least 1000 miles on a rubber life raft to a tiny South Sea isle and later were picked up by Navy craft. They now are at Pearl Harbor and virtually recovered. The Navy announced their rescue last month.

Their story is one of quick thinking, resourcefulness and raw courage. Dixon now is partially deaf. When an albatross alighted unexpectedly on the raft, Aldrich fired a pistol too close to Dixon's ear. Dixon jumped into the sea to retrieve the bird and its flesh helped keep the men alive. It was on the night of January 16 that Dixon decided he had better set his plane down on the sea while enough gas remained to keep the plane under control. The plane sank quicker than expected and the men were unable to save the rations, water or instruments. While floundering in the water, they managed to inflate the rubber life raft, its inside dimensions only 28" x 80". In this precarious haven the three crowded for the perilous days ahead.

"We were not concerned much the first night," Dixon said. "We knew searchers would be out and sure enough we saw a plane in the morning; but it did not see us and that was the last we saw of possible help." The raft was not equipped with oars and for some time the trio were just drifting. Later Dixon's shoes were rigged so that they could be used as hand-oars. "Don't say we drifted to the island—we worked," Dixon said. He told of shooting a big leopard shark which came so close he merely held the pistol against the shark's head. He related that still later he struck another shark with his bare fists. The men lost from 20 to 50 pounds each in the long ordeal.

They rigged a sea anchor from a life jacket, using it whenever winds were adverse. They used their own underclothing as sponges when it rained. After the cloth was saturated, they wrung it out to fill a small container with the precious water. At all times they knew where they wanted to go—toward islands to the south. The sea was smooth durin[g] the first part of the trip. Later, when they finally sighted land, they were in the middle of one of the worst hurricanes in South Sea history. Time and again they were tossed into the ocean, only to climb back. The last few days they did not have a stitch of clothing and were sunburned so badly it took weeks to heal. Their total food hardly would make more than one good meal, but Dixon insisted coffee and cigarettes were missed more than the food. Aldrich proved the best fisherman, but once his hand was bitten by a shark. They had practically no water the first four days, until they figured out the idea of using their underclothing as a sponge. Neither did they have food. Then Aldrich stabbed a fish with his pocketknife and they chewed the raw flesh. They said it was like chewing rubber. The albatross was next on the menu. After part had been eaten Dixon noted at night that the bird seemed to emit light. He feared it had become poisonous fro[m] feeding in phosphorescent waters and heaved the remainder of the carcass overboard.

On the 7th day Aldrich stabbed a 4-foot shark. "We ripped the shark open with pliers," Dixon said. "We had read about vitamins in sharks' livers so ate that first. It was ver[y] good. We also ate two sardines found in the shark's stomach with a great deal of relish. And the next day we polished off the shark itself. On the fourteenth day Aldrich caught another fish and I caught a bird by reaching up for it when it landed on the raft. It was some sort of a tern, but to us tasted like fried chicken. We devoured it on the spot. On the 21st day we picked up a coconut. Things looked pretty bad the thirty-third day. We had lost our clothes, all but one shoe, and the pistol long since had corroded into uselessness. We found ourselves periodically in a daze."

"I was lying down as best we could the morning of the [illegible] day when Gene (Aldrich's nickname) said he saw corn. I didn't pay any attention. [illegible] We tried to paddle all day and finally staggered ashore [illegible] with the hurricane just getting to top force. We decided when we came ashore we would try to march and if the island was Japanese, they would not find us crawling up to them. But we could not straighten up. The next morning, island natives on beach patrol saw us, gave us kernals of coconuts and sent for help. They carried us to the governor's house."

Dixon finished his recital with the words: "These boys showed plenty of stuff all the way through." Asked how he [illegible] "Regularly."

Sample Pilot History Card

NAME AND RANK 3689 DIXON, Harold F., ACMM(NAP)
SERVICE AND CORPS USN
ORGANIZATION VT-6 USS ENTERPRISE
DATE OF PILOT RATING — FLYING EXPERIENCE (YEARS) 10-15
HRS. THIS MODEL LAST 3 MOS 59.5 TOTAL HOURS 4300.2
PREVIOUS ACCIDENT RECORD 1934
Injury: Missing

PERSONNEL

NAME AND RANK	DUTY	INJ
Anthony J. Pastula, AOM2c, USN (Missing)	PO	8
Gene D. Aldrich, RM3c, USN (Missing)	PO	8

AIRPLANE CLASS VTB (L)
MODEL AND NUMBER TBD-1 #0335

REMARKS: Missing. Personnel found after 34 days in rubber boat [illegible]

NATURE OF ACCIDENT
PERCENTAGE EACH CAUSE
100% Misc. (war conditions) 50% Undetermined

DATE OCCURRED 16 January 1942 1915-2000 NIGHT
LOCATION At sea
DETAILED CAUSE
FLYING OR LANDING CONDITIONS Average. Visibility good. Wind moderate, scattered squalls.

ANALYSIS

Dixon took off at 1330 on a routine scouting flight to search to a distance of 175 miles a 10° relative sector. His plane was sighted by the pilot of another plane engaged in the search at the outer extremity of his sector at a distance of approximately 75 miles from his (Dixon's) proper position. He was due back over the ship at 1805. At 1900 very faint radio communication was established between the ship and plane. Dixon reported he had 20 gallons of fuel remaining. Reception was very poor and his position could not be ascertained. At 1915 he reported he was very low on fuel. No further contact was made and the plane did not return. A search was made the following day covering the largest area practicable commensurate with mission of force. No trace of aircraft or personnel was found.

Purpose of flight: Scouting - J

Entered on Pilot's Card

CONFIDENTIAL

U.S. NAVY BUREAU OF AERONAUTICS AIRCRAFT TROUBLE ANALYSIS

Example of Single Aircraft Record Card

The Naval Aviation History Branch will also provide researchers with record cards of a single aircraft at no charge. You must provide them with the aircraft type and bureau number (Navy serial number).

Reel No.	Microdex No.	Model	Crashes Listed By Date
14	3	AD	Apr 1947 thru Dec 1949
14	4	AD	Jan 1950 thru Dec 1950
14	5	AD	Jan 1951 thru Jul 1951
14	6	AD	Aug 1951 thru Dec 1951
14	7	AD	Jan 1952 thru Jun 1952
14	8	AD	Jun 1950 thru Jun 1952
16	1	AE	Jan 1950 thru Dec 1950
23	4	AJ	Jan 1948 thru Dec 1948
23	5	AM	Jan 1947 thru Dec 1947
23	6	AV	Feb 1952 thru Jun 1952
23	7	BD	Jan 1942 thru Dec 1942
16	2	BD-2	Jan 1942 thru Dec 1944
23	8	BT	Jan 1941 thru Dec 1941
16	2	BTD-1	Jan 1945 thru Dec 1952
16	3	FD-1	Jan 1947 thru Dec 1947
28	2	Fatalities	Sep 1941 thru Jul 1951
23	9	FF	Jan 1942 thru Dec 1942
12	8	FG	Jan 1943 thru Dec 1944
12	9	FG	Jan 1945 thru Mar 1945
13	1	FG	Apr 1945 thru Jul 1945
13	2	FG	Aug 1945 thru Aug 1946
13	3	FG	Sep 1946 thru Jul 1947
13	4	FG	Aug 1948 thru Dec 1951
13	5	FG	Jan 1952 thru Jun 1952
24	1	FH	Nov 1947 thru Mar 1952
23	3	FJ	Jan 1947 thru Dec 1947
15	1	FM	Dec 1942 thru Dec 1943
15	2	FM	Jan 1944 thru Apr 1944
15	3	FM	May 1944 thru Jun 1944
15	4	FM	Jul 1944 thru Aug 1944
15	5	FM	Sep 1944 thru Oct 1944
15	6	FM	Nov 1944 thru Dec 1944
15	7	FM	Jan 1945
15	8	FM	Feb 1945
15	9	FM	Mar 1945
17	10	FM	Apr 1945 thru Jun 1945
18	1	FM	Jul 1945 thru Nov 1946
16	3	FR	Jan 1944 thru Dec 1949
23	10	FR	Jan 1945 thru Jun 1947
11	7	F2F	Jul 1941 thru Nov 1941
16	3	F2G-1	Jan 1945 thru Dec 1946
13	6	F2H	Jan 1948 thru Dec 1948
16	3	F2L-1	Jan 1946 thru Dec 1946
16	3	F2T	Jan 1945 thru Dec 1946
18	2	F3A	Jul 1941 thru Dec 1941
18	3	F3A	Jul 1943 thru Oct 1946
13	7	F3D	Dec 1950 thru Apr 1952

Reel No.	Microdex No.	Model	Crashes Listed By Date
13	8	F3F	Jul 1941 thru Jan 1943
16	3	F4B	Jan 1941 thru Dec 1942
13	9	F4F	Jul 1941 thru Oct 1942
13	10	F4F	Nov 1942 thru Dec 1943
14	1	F4F	Jan 1944 thru Aug 1945
14	2	F5F	Feb 1942 thru Jun 1942
9	1	F4U	Aug 1942 thru Dec 1943
9	2	F4U	Jan 1944 thru Mar 1944
9	3	F4U	Apr 1944 thru Sep 1944
9	4	F4U	Oct 1944 thru Dec 1944
9	5	F4U	Jan 1945 thru Mar 1945
9	6	F4U	Apr 1945 thru May 1945
9	7	F4U	Jun 1945
9	8	F4U	Jul 1945 thru Dec 1945
9	9	F4U	Jan 1946 thru Jun 1946
10	1	F4U	Jun 1946 thru Dec 1946
10	2	F4U	Jan 1947 thru Aug 1947
10	3	F4U	Sep 1947 thru Apr 1948
10	4	F4U	May 1948 thru Dec 1948
10	5	F4U	Jan 1948 thru May 1949
10	6	F4U	Jun 1949 thru Sep 1949
10	7	F4U	Oct 1949 thru Dec 1949
10	8	F4U	Jan 1950 thru Aug 1950
10	9	F4U	Sep 1950 thru Dec 1950
11	1	F4U	Jan 1951 thru Apr 1951
11	2	F4U	May 1951 thru Jun 1951
11	3	F4U	Jul 1951 thru Oct 1951
11	4	F4U	Nov 1951 thru Dec 1951
11	5	F4U	Jan 1952 thru Feb 1952
11	6	F4U	Mar 1952 thru Jun 1952
6	1	F6F	Dec 1942 thru Dec 1943
6	2	F6F	Jan 1944 thru Feb 1944
6	3	F6F	Mar 1944 thru Apr 1944
6	4	F6F	May 1944 thru Jun 1944
6	5	F6F	Jul 1944
6	6	F6F	Aug 1944
6	7	F6F	Sep 1944
6	8	F6F	Oct 1944
6	9	F6F	Nov 1944
7	1	F6F	Dec 1944
7	2	F6F	Jan 1945
7	3	F6F	Feb 1945
7	4	F6F	Mar 1945
7	5	F6F	Apr 1945
7	6	F6F	May 1945
7	7	F6F	Jun 1945
7	8	F6F	Jul 1945

Reel No.	Microdex No.	Model	Crashes Listed By Date	Reel No.	Microdex No.	Model	Crashes Listed By Date
7	9	F6F	Aug 1945	28	5	N2S	Sep 1943 thru May 1944
8	1	F6F	Sep 1945 thru Dec 1945	28	6	N2S	Jun1944 thru Jan 1945
8	2	F6F	Jan 1946 thru Jul 1946	28	7	N2S	Feb 1945 thru May 1945
8	3	F6F	Jul 1946 thru Dec 1946	28	8	N2S	Jun 1945 thru Apr 1946
8	4	F6F	Jan 1947 thru Dec 1947	28	9	N2S	May 1946 thru Jun 1952
8	5	F6F	Jan 1948 thru Dec 1948	26	6	N2T	May 1942 thru Aug 1944
8	6	F6F	Jan 1949 thru Dec 1949	26	5	N3N	Jul 1941 thru Jun 1952
8	7	F6F	Jan 1950 thru Dec 1950	17	3	OBM	Jan 1947 thru Jun 1952
8	8	F6F	Jan 1951 thru Dec 1951	27	4	OE	Dec 1951 thru May 1952
8	9	F6F	Jan 1952 thru Jun 1952	27	7	OS2N	Mar 1942 thru Sep 1945
16	3	F6U	Jan 1947 thru Dec 1950	27	5	OS2U	Jul 1941 thru Sep 1947
11	8	F7F	Jan 1944 thru Dec 1944	27	3	OY	Oct 1943 thru Nov 1951
11	9	F7F	Jul 1944 thru Dec 1948	27	6	O3U	Aug 1941 thru Jan 1943
11	10	F7F	Jan 1949 thru Sep 1950	16	6	PB	Jan 1948 thru Dec 1951
12	1	F8F	Mar 1945 thru Jun 1947	25	5	PBJ	Mar 1943 thru Oct 1947
12	2	F8F	Jul 1947 thru Dec 1947	17	1	PBM	Oct 1941 thru Dec 1944
12	3	F8F	Jan 1948 thru Dec 1948	17	2	PBM	Jan 1945 thru Dec 1946
12	4	F8F	Jan 1949 thru Dec 1949	25	4	PBO	Nov 1941 thru Feb 1944
12	5	F8F	Jan 1950 thru Dec 1950	25	6	PBY	Jul 1941 thru Dec 1943
12	6	F8F	Jan 1951 thru Dec 1951	25	8	PB2B	Feb 1945 thru May 1946
12	7	F8F	Jan 1952 thru Jun 1952	28	1	PB2Y	May 1942 thru Jan 1946
23	1	F9F	Oct 1948 thru Apr 1951	25	9	PB4Y	Feb 1943 thru Mar 1945
23	2	F9F	May 1951 thru Jun 1952	25	10	PB4Y	Apr 1945 thru Sep 1945
24	2	GB	Aug 1941 thru Jun 1946	27	8	PV	Oct 1942 thru Aug 1951
24	3	GH	Apr 1942 thru Aug 1946	26	1	P2V	Jun 1945 thru Jun 1952
24	4	Helos	Oct 1943 thru Jun 1952	27	8	P2V	Aug 1951
25	2	JD	Mar 1946 thru May 1952	28	1	P2Y	Nov 1941 thru Feb 1942
16	4	JF	Jan 1941 thru Dec 1943	27	9	P4M	Feb 1950 thru Feb 1952
24	10	JM	Oct 1943 thru Jul 1947	27	10	P4Y-2	Aug 1951 thru Jun 1952
25	1	JRB	Oct 1941 thru Jun 1952	16	7	RB	Jan 1944 thru Dec 1944
16	4	JRC	Jan 1943 thru Dec 1945	16	7	R2D	Jan 1941 thru Dec 1943
24	9	JRF	Jul 1941 thru Sep 1956	17	4	R4D	Jul 1942 thru Jun 1952
24	7	JRM	Aug 1945 thru Aug 1950	17	5	R4Q	Feb 1951 thru Jun 1952
24	8	JRS	Jul 1941 thru Jun 1944	17	7	R5C	Mar 1943 thru Jun 1952
24	6	J2F	Jul 1941 thru Oct 1948	17	6	R5D	Mar 1943 thru Dec 1943
24	5	J4F	Oct 1942 thru Jun 1948	17	8	R5D	Feb 1944 thru Jun 1952
25	3	LTA	Jun 1942 thru Apr 1952	17	9	SBC	Jul 1941 thru Dec 1941
16	10	Misc.	Jan 1941 thru Dec 1952	22	1	SBD	Jul 1941 thru Dec 1942
27	2	NE	Apr 1942 thru Jun 1946	22	2	SBD	Jan 1943 thru Jul 1943
27	1	NH	May 1943 thru Nov 1946	22	3	SBD	Aug 1943 thru Oct 1943
26	9	NJ	Oct 1941 thru Jun 1944	22	4	SBD	Nov 1943 thru Dec 1943
26	10	NP	Dec 1941 thru Jun 1943	22	5	SBD	Jan 1944 thru Feb 1944
16	5	NR	Jan 1942 thru Dec 1944	22	6	SBD	Mar 1944 thru Apr 1944
26	8	NR	Aug 1941 thru May 1943	22	7	SBD	May 1944 thru Jun 1944
26	7	NS	Oct 1941 thru Jan 1943	32	7	SBD	Jul 1944 thru Sep 1944
28	3	N2S	Jul 1941 thru Feb 1943	32	8	SBD	Oct 1944 thru Dec 1944
28	4	N2S	Mar 1943 thru Aug 1943	32	9	SBD	Jan 1945 thru Jul 1945

Reel No.	Microdex No.	Model	Crashes Listed By Date
32	10	SBD	Aug 1945 thru Mar 1946
34	2	SBF	Aug 1944 thru Jan 1946
33	10	SBN	Jul 1941 thru Feb 1943
34	1	SBU	Jul 1941 thru Jul 1943
33	7	SBW	Jan 1944 thru Sep 1946
33	8	SBW	Oct 1946 thru Sep 1948
30	3	SB2A	Feb 1943 thru Aug 1944
29	4	SB2C	Nov 1942 thru May 1944
29	5	SB2C	Jun 1944 thru Aug 1944
29	6	SB2C	Sep 1944 thru Nov 1944
29	7	SB2C	Dec 1944 thru Mar 1945
29	8	SB2C	Apr 1945 thru Jun 1945
29	9	SB2C	Jul 1945 thru Sep 1945
29	10	SB2C	Oct 1945 thru Dec 1946
30	1	SB2C	Jan 1947 thru Oct 1947
30	2	SB2C	Nov 1947 thru Oct 1951
33	9	SB2U	Jul 1941 thru Jul 1943
32	3	SC	Sep 1944 thru Jan 1949
16	8	SF-1	Jan 1942 thru Dec 1942
26	2	SNB	Nov 1942 thru Dec 1945
26	3	SNB	Jan 1946 thru Jun 1950
26	4	SNB	Jul 1950 thru Jun 1952
32	4	SNC	Aug 1941 thru Sep 1944
19	1	SNJ	Jul 1941 thru Jan 1943
19	2	SNJ	Feb 1943 thru Jul 1943
19	3	SNJ	Aug 1943 thru Dec 1943
19	4	SNJ	Jan 1944 thru Mar 1944
19	5	SNJ	Apr 1944 thru Jul 1944
19	6	SNJ	Aug 1944 thru Oct 1944
19	7	SNJ	Nov 1944 thru Dec 1944
19	8	SNJ	Jan 1945 thru May 1945
19	9	SNJ	Jun 1945 thru Aug 1945
20	1	SNJ	Sep 1945 thru Dec 1945
20	2	SNJ	Jan 1946 thru Jun 1946
20	3	SNJ	Jul 1946 thru Dec 1946
20	4	SNJ	Jan 1947 thru May 1947
20	5	SNJ	Jun 1947 thru Dec 1947
20	6	SNJ	Jan 1948 thru Apr 1948
20	7	SNJ	May 1948 thru Sep 1948
20	8	SNJ	Oct 1948 thru Dec 1948
20	9	SNJ	Jan 1949 thru Dec 1949
21	1	SNJ	Jan 1950 thru Dec 1950
21	2	SNJ	Jan 1951 thru Sep 1951
21	3	SNJ	Oct 1951 thru Dec 1951
21	4	SNJ	Jan 1952 thru Jun 1952
32	5	SNV	Aug 1941 thru May 1944
32	6	SNV	Jun 1944 thru Oct 1945

Reel No.	Microdex No.	Model	Crashes Listed By Date
29	3	SOC	Jul 1941 thru Sep 1946
28	10	SON	Jul 1941 thru Nov 1944
29	1	SU	Aug 1941 thru Mar 1943
29	2	SO3C	May 1942 thru Aug 1945
33	1	TBD	Jul 1941 thru Sep 1943
32	2	TBF	Jul 1943 thru Jun 1944
33	2	TBF	Jul 1941 thru Aug 1943
33	3	TBF	Sep 1943 thru Feb 1944
33	4	TBF	Mar 1944 thru Sep 1944
33	5	TBF	Oct 1944 thru May 1945
33	6	TBF	Jun 1945 thru Dec 1945
30	4	TBM	Apr 1943 thru Apr 1944
30	5	TBM	May 1944 thru Sep 1944
30	6	TBM	Oct 1944 thru Dec 1944
30	7	TBM	Jan 1945 thru Feb 1945
30	8	TBM	Mar 1945 thru May 1945
30	9	TBM	Jun 1945
30	10	TBM	Jul 1945
31	1	TBM	Aug 1945 thru Dec 1945
31	2	TBM	Jan 1946 thru Jan 1947
31	3	TBM	Feb 1947 thru Apr 1948
31	4	TBM	May 1948 thru Jul 1949
31	5	TBM	Aug 1949 thru Dec 1950
31	6	TBM	Jan 1951 thru Dec 1951
31	7	TBM	Jan 1952 thru Jun 1952
16	9	TBY	Jan 1944 thru Dec 1945
31	8	TBY	Jul 1945 thru Jun 1947
16	9	TDC	Jan 1943 thru Dec 1944
16	9	TDN	Jan 1943 thru Dec 1943
31	9	TD2C	Jul 1944 thru Mar 1950
31	10	TO	Oct 1948 thru Mar 1952
32	1	TV	Mar 1952 thru Jun 1952

APPENDIX III

Obtaining U.S. Air Force aircraft history cards

The U.S. Air Force will also provide researchers with the individual record card for a certain aircraft at no charge, or you can order an entire reel of aircraft record cards for $20. To order a record card or a microfilm reel, contact:

Department of the Air Force
USAF Historical Research Center/HD
Maxwell Air Force Base, Ala. 36112-6678

U.S. Air Force Aircraft Record Cards Available On 16mm Microfilm

ACR-1	216	thru	623
ACR-2	24-216	thru	26-423
ACR-3	26-424	thru	28-213
ACR-4	28-214	thru	29-180
ACR-5	29-182	thru	30-209
ACR-6	30-210	thru	31-86
ACR-7	31-87	thru	31-405
ACR-8	31-406	thru	32-169
ACR-9	32-170	thru	33-277
ACR-10	33-278	thru	08318
ACR-11	08320	thru	KF232
ACR-12	KF233	thru	Z7499
ACR-13	36-1	thru	36-446
ACR-14	37-1	thru	37-520
ACR-15	37-521	thru	37-679
ACR-16	38-1	thru	38-513
ACR-17	38-514	thru	38-610
ACR-18	39-1	thru	40-818
ACR-19	40-819	thru	40-1638
ACR-20	40-1639	thru	40-2731
ACR-21	40-2732	thru	41-842
ACR-22	41-843	thru	41-2407
ACR-23	41-2408	thru	41-6039
ACR-24	41-6040	thru	41-8100
ACR-25	41-8101	thru	41-9465
ACR-26	41-9468	thru	41-11631
ACR-27	41-11632	thru	41-13472
ACR-28	41-13473	thru	41-15706
ACR-29	41-15707	thru	41-17845
ACR-30	41-17846	thru	41-20343
ACR-31	41-20344	thru	41-22467
ACR-32	41-22468	thru	41-24808
ACR-33	41-24809	thru	41-27325
ACR-34	41-27326	thru	41-30049
ACR-35	41-30050	thru	41-32171
ACR-36	41-32172	thru	41-35037
ACR-37	41-35038	thru	41-36952
ACR-38	41-36953	thru	41-37199
ACR-39	41-37200	thru	41-39600
ACR-40	42-1	thru	42-2602
ACR-41	42-2607	thru	42-4641
ACR-42	42-4642	thru	42-6958
ACR-43	42-6959	thru	42-9204
ACR-44	42-9205	thru	42-12774
ACR-45	42-12775	thru	42-15274
ACR-46	42-15275	thru	42-17279
ACR-47	42-17280	thru	42-19554
ACR-48	42-19555	thru	42-23032
ACR-49	42-23033	thru	42-25446
ACR-50	42-25447	thru	42-28121
ACR-51	42-28122	thru	42-30426
ACR-52	42-30427	thru	42-32912
ACR-53	42-32913	thru	42-36065
ACR-54	42-36066	thru	42-38804
ACR-55	42-38805	thru	42-40811
ACR-56	42-40812	thru	42-43283
ACR-57	42-43284	thru	42-46433
ACR-58	42-46434	thru	42-50040
ACR-59	42-50041	thru	42-52579
ACR-60	42-52580	thru	42-55761

U.S. Air Force Aircraft Record Cards Available On 16mm Microfilm (cont'd)

ACR-61	42-55762	thru	42-58763
ACR-62	42-58764	thru	42-64188
ACR-63	42-64189	thru	42-67129
ACR-64	42-67130	thru	42-69620
ACR-65	42-69621	thru	42-72221
ACR-66	42-72222	thru	42-74600
ACR-67	42-74601	thru	42-77622
ACR-68	42-77623	thru	42-79972
ACR-69	42-79973	thru	42-84905
ACR-70	42-84906	thru	42-88654
ACR-71	42-88655	thru	42-92669
ACR-72	42-92670	thru	42-95390
ACR-73	42-95391	thru	42-98074
ACR-74	42-98075	thru	42-100575
ACR-75	42-100576	thru	42-104107
ACR-76	42-104108	thru	42-106379
ACR-77	42-106380	thru	43-356
ACR-78	43-357	thru	43-2321
ACR-79	43-2322	thru	43-4990
ACR-80	43-4991	thru	43-7345
ACR-81	43-7346	thru	43-10059
ACR-82	43-10060	thru	43-13176
ACR-83	43-13177	thru	43-21770
ACR-84	43-21771	thru	43-24170
ACR-85	43-24171	thru	43-26954
ACR-86	43-26955	thru	43-29253
ACR-87	43-29255	thru	43-32464
ACR-88	43-32465	thru	43-36691
ACR-89	43-36692	thru	43-39253
ACR-90	43-39254	thru	43-41627
ACR-91	43-41628	thru	43-45476
ACR-92	43-45477	thru	43-50199
ACR-93	43-50200	thru	43-52437
ACR-94	44-1	thru	44-3969
ACR-95	44-3970	thru	44-7450
ACR-96	44-7451	thru	44-10951
ACR-97	44-10952	thru	44-13309
ACR-98	44-13310	thru	44-17058
ACR-99	44-17059	thru	44-20809
ACR-100	44-20810	thru	44-24925
ACR-101	44-24926	thru	44-26199
ACR-102	44-26200	thru	44-30562
ACR-103	44-30563	thru	44-33272
ACR-104	44-33273	thru	44-40800
ACR-105	44-40801	thru	44-44399
ACR-106	44-44400	thru	44-50636
ACR-107	44-50637	thru	44-63538
ACR-108	44-63539	thru	44-70483
ACR-109	44-70484	thru	44-73698
ACR-110	44-73700	thru	44-77739
ACR-111	44-77740	thru	44-80971
ACR-112	44-80975	thru	44-84873
ACR-113	44-84878	thru	44-88548
ACR-114	44-88550	thru	45-5855
ACR-115	45-5856	thru	45-14629
ACR-116	45-14630	thru	45-58850
ACR-117	45-58851	thru	49-784
ACR-118	49-794	thru	51-600

The following records were out of file at the time the ACR rolls were filmed:

OACR-1	41-32145	thru	50-1639
OACR-2	41-327	thru	42-44668
OACR-3	42-44670	thru	44-35689
OACR-4	44-35705	thru	44-83699
OACR-5	44-83723	thru	45-34995
OACR-6	45-34988	thru	49-2426
OACR-7	49-2427	thru	British planes
OACR-8	41-1143	thru	44-85467
OACR-9	44-85531	thru	51-10909
OACR-10	51-10910	thru	AFF+NGG losses, Navy planes

APPENDIX IV

Obtaining the right USGS map

The United States Geological Survey publishes topographical maps for every inch of the United States. The maps range in cost from a minimum of $2.50 to as high as $10 and are available at many bookstores, blueprint stores, engineering supply stores, main libraries, and university libraries. Consult your local phone book.

To order USGS maps, you must first obtain the following items for the state in which your wreck is located:

- Δ **Catalog of Topographical and Other Published Maps**
- Δ **Index to Topographic and Other Map Coverage**

Both publications are free of charge. To order the above publications, write to:

Map Distribution/Sales
U.S. Geological Survey
Box 25286, Federal Center
Denver, Colo. 80225

The Catalog of Topographical and Other Published Maps will provide you with a list of USGS map dealers in your state. Once you have selected your map, you can order it by mail or visit one of the retailers listed in the map catalog. Be sure to pick up the Guide to Topographic Map Symbols.

Maps must be ordered by map name, state, and series/scale. Payment should be made by check or money order made payable to the **Department of the Interior — USGS.**

APPENDIX V

Individuals and groups involved in aviation archaeology

Gary Larkins P.O. Box 7108 Auburn, Calif.	Mr. Larkins has recovered a number of rare aircraft from remote locations. He will pay a finders fee for certain types of aircraft wrecks in fresh water lakes.
Sig Unander Air Art Northwest 846 South Ginger St. Cornelius, Ore. 97113 (503) 359-0424	Mr. Unander is the author of *The Master File: World War II and Korean War-Era Aircraft Losses in the Northwest United States*, a well-researched book detailing more than 270 aircraft losses in the Northwestern United States.
Robert Kropp 208 N. MacDonald Ave. Mesa, Ariz. (602) 644-1581	Mr. Kropp has recovered a number of World War II fighters and trainers. He specializes in aircraft in the Southwest.
Jay Wisler 2710 Clark Road Tampa, Fla. 33618 (813) 968-5048	Mr. Wisler is interested in aircraft wrecks in the Southern states — Florida in particular. He also buys and sells World War II aircraft parts.
Patrick Macha 17111 Camelot Circle Huntington Beach, Calif. 92649	Mr. Macha is the author of *Aircraft Wrecks in the Mountains and Deserts of California.*
Redwood Empire Aviation Historical Society P.O. Box 9073 Santa Rosa, Calif. 95405	The Redwood Empire Aviation Historical Society is involved in preserving aviation history in the Northern California coastal region. Members of the society have visited a number of interesting wreck sites.
TIGHAR 2812 Fawkes Drive Wilmington, Del. 19808 (302) 994-4410	The International Group for Historic Aircraft Recovery.
In Flight Aviation News P.O. Box 620447 Woodside, Calif. 94062 (415) 364-8110	*In Flight Aviation News* often runs articles on historic aircraft crashes.

ABOUT THE AUTHOR

Nicholas A. Veronico is an aviation historian specializing in U.S. Naval aircraft of World War II and Korea. He is also interested in World War II aircraft armament and aircraft crash sites. He has written more than 500 aviation-related newspaper and magazine articles, as well as three Air Force Base museum guide books. Nicholas A. Veronico serves as the managing editor of *In Flight Aviation News.*